PRACTICAL EMULSIONS

PRACTICAL EMULSIONS

Volume II

Applications

H. Bennett
Technical Director
B.R.Laboratory
Miami Beach, Florida

Jack L. Bishop Jr., Ch.E.
Dow Corning Corporation
Midland, Michigan

Max F. Wulfinghoff, M.E.
Tampa, Florida

CHEMICAL PUBLISHING COMPANY, INC.
New York

Third Edition

Printed in the United States of America

Foreword

The first volume of *Practical Emulsions* has attempted to convey a general impression of the science of the formulation and manufacture of emulsions. Salient points have been presented in as simple a manner as possible, while detailed examination of specific points of the theory has been left to those whose qualifications and interests better fit them for such tasks.

In this volume, I have compiled basic formulations of many types of emulsion. This compilation is intended to provide graphic illustration of the breadth of emulsion technology. The formulations can also provide a starting point for the development of new and improved emulsion products and represent the types of product that can be made and their major constituents.

These formulations are provided by manufacturers to illustrate the use of their products. The suitability of any product for marketing, or the determination of the patent position regarding any formulation, is the responsibility of the person who manufactures and/or sells the product.

Many firms have generously provided formulation suggestions, and they merit my sincere appreciation for their help. Their contributions are acknowledged by appropriate references at the end of each chapter.

July 1967

J. L. Bishop, Jr.

Contents

chapter 1

AGRICULTURAL EMULSIONS

The formulation of an agricultural emulsion requires unique considerations. The emulsion must possess sufficient stability to carry the active ingredient to the surface of the leaf, but must not allow re-emulsification of the active ingredients. The typical agricultural emulsion is comprised of a concentrate that contains the active ingredient (pesticide, herbicide, etc.), a surface-active agent, and a solvent. The emulsion is formed when the concentrate is mixed with water immediately prior to use. Therefore, the concentrate must immediately form an emulsion upon dilution with water.

The surfactant in a typical agricultural emulsion should provide an emulsion of small particle size. The surfactant should also aid wetting of the leaves by the spray. The surfactant can not permit re-emulsification of the spray on the leaf by rain water. Some investigators have found that anionic–nonionic blends of surfactants are more effective at a lower concentration and cost than unblended nonionic surfactants.[1] Cationic surfactants are often used as emulsifying and wetting agents in agricultural emulsions. If the spray contains a cationic rather than a nonionic wetting agent, more of the oil phase will be deposited on the leaves. The type and concentration of the cationic surfactant also influences the deposition of the oil phase on the leaves. Generally, the higher the efficiency of the emulsifying agent the lower the deposition of the oil phase on the leaf.[2]

A solvent is often found in the oil phase of an agricultural emulsion and is used to keep the active ingredients (pesticide, herbicide, etc.) and emulsifying agent in solution. The solvent should have a high flash point to avoid shipping restrictions.[3]

The adhesion of the oil-phase droplet to the leaf is affected by mechanical action as well as by the type and amount of surfactant and solvent. The size of the droplet and its velocity affect the amount of the oil phase that will remain on the leaf. The evaporation rate of the emulsion is often lowered by the addition of saturated fatty acids to change the retention characteristics of the oil droplets.[4]

Agricultural sprays that are applied from low flying aircraft create special formulation problems. Aerial application of pesticides or herbicides is greatly affected by wind induced spray drifting. Although mechanical means of reducing drifting are occasionally used, most aerial agricultural sprays are water-in-oil emulsions rather oil-in-water emulsions. These "invert" emulsions form large droplets that are less affected by wind drifting.[5]

Aldrin Formulations

Formula No. 1[6]

Aldrin	26.8*
Xylene	31.6
Kerosene	31.6
Alkapent TD 100	4.0
Alkapent D NP 100	6.0

2 lb/gal

Chlordane Formulations

No. 2[6]

Chlordane	72.0
Kerosene	20.0
Alkapent TD 100	3.0
Alkapent D NP 100	5.0

8 lb/gal

No. 3[1]

	Soft water	*Hard water*
Chlordane	45.0	45.0
Triton X-151		1.3
Triton X-171	4.0	2.7
Kerosene	51.0	51.0

8.9 lb/gal

DDT Formulations

No. 4[8]

DDT	25.0
Dowfax 9N9	5.0
Xylene	70.0

No. 5[9]

	Soft water	*Hard water*
DDT	25.0	25.0
Triton X-151	0.9	2.5
Triton X-171	2.1	0.5
Xylene	72.0	72.0

8.0 lb/gal

* Throughout this volume all figures are parts by weight, except where otherwise indicated.

Dieldrin Formulations

No. 6[10]

Dieldrin	20.0
Dowfax 9N9	5.0
Xylene	75.0

No. 7[11]

Dieldrin	19.0
Heavy aromatic Naphtha	77.0
Alkapent TD 100	4.0
Alkapent D NP 100	6.0

1.5 lb/gal

Lindane Formulations

No. 8[12]

Lindane	20.0
Dowfax 9N9	5.0
Kerosene	75.0

No. 9[13]

Lindane	20.0
Emulsifier C	7.5
Velsicol AR-50	32.5
Isophorone	40.0

No. 10[14]

Malathion Wettable Powder

Malathion (97% active)	25.8
Celite 209	28.7
Barden clay	42.5
Sodium lignosulfonate	2.0
Dowfax 9N9	1.0

No. 11[15]

Malathion–Xylene

	Soft water	*Hard water*
Malathion (95%)	52.7	52.7
Triton X-152		4.0
Triton X-172	5.0	1.0
Xylene	42.3	42.3

8.6 lb/gal

No. 12[16]

Methyl Parathion Emulsion Concentrate

Technical Methyl Parathion	31.40
Xylene	63.60
Atlox 3335	4.25
Atlox 8916P	0.75

8.05 lb/gal

No. 13[17]

Ronnel Formulation

0, 0-Dimethyl, 0-2, 4, 5-Tricholrophenyl Phosphorothioate (*Ronnel*)	24.0
Dowfax 9N9	2.5
Petroleum Sulfonate	2.5
Xylene	71.0

No. 14[17]

Toxaphene Formulation

Toxaphene	60.0
Dowfax 9N9	5.0
Kerosene	35.0

No. 15[17]

Toxaphene–DDT Formulation

Toxaphene	40.0
DDT	20.0
Dowfax 9N9	5.0
Xylene	35.0

2, 4-D Formulations

No. 16[17]

Isooctyl Ester of 2, 4-D Formulation

Isooctyl Ester of 2, 4-D	65.0

Dowfax 9N9	2.2
Petroleum Sulfonate	1.8
Heavy Aromatic Naphtha	31.0

No. 17[17]

Butyl Ester of 2, 4-D Formulations

Butyl Ester of 2, 4-D	56.0
Dowfax 9N9	2.4
Petroleum Sulfonate	1.6
Heavy Aromatic Naphtha	40.0

No. 18[18]

	Soft water	*Hard water*
2,4-D Butyl Ester	57.0	57.0
Triton X-152	0.6	2.0
Triton X-172	3.4	2.0
HAN-132	39.0	39.0

8.95 lb/gal

No. 19[18]

2,4,5-T Isooctyl Ester

	Soft water	*Hard water*
2,4,5-T Isooctyl Ester	65.8	65.8
Triton X-152	0.4	1.6
Triton X-172	3.6	2.4
Diesel Oil	30.2	30.2

8.82 lb/gal

No. 20[19]

2-Ethylhexyl-2, 4, 5-Trichlorophenoxyacetate

2-Ethylhexyl-2, 4, 5-Trichlorophenoxyacetate	64.2
Monazoline O	5.3
Monapal T	2.7
Xylene	27.8

Fruit Coating Wax Emulsions

No. 21[20]

Caustic Soda	6
Triethanolamine	20
Stearic Acid	42
Paraffin Wax	165
Carnauba Wax	55
Shellac	100
Water	2000

Dilute before use

No. 22[21]

Paraffin Wax	553
Carnauba Wax	68
Cottonseed Oil	98
Oleic Acid	183
Triethanolamine	98
Water (containing Soda Ash)	*q.s.*

No. 23[22]

Paraffin Wax	168.0
Beeswax	42.0
Oleic Acid	22.0
Sodium Bicarbonate	6.6
Salt	2.2
Water	599.2

CATTLE DIPS

Toxaphene Formulations

No. 24[23]

Toxaphene (90%)	66.6
Kerosene	18.4
Atlox 3404	6.0
Atlox 3403	9.0

No. 25[23]

Toxaphene (90%)	55.3
Malathion (95%)	5.3
XA	24.1
Atlox 3404	6.0
Atlox 3403	9.0

No. 26[23]

Toxaphene (90%)	55.6
Lindane	2.0
XA	27.4
Atlox 3404	6.0
Atlox 3403	9.0

No. 27[23]

Soil Fumigant*

Nemogon® (3)	83.7
Xylene	11.3
Atlox 3404	1.0
Atlox 3403	4.0

No. 28[23]

Miticide*

Aramite (93%)	91.0
XA	2.0
Atlox 3409	7.0

* For longer shelf life, add 0.5% Epichlorhydrin to these formulas.

References

1. J. K. EATON, "General Requirements of Pesticide Emulsions," *Chem. Ind.* Nov. 10, 1962, pp. 1914-6.
2. C. G. L. FARMIDGE, "Retention of Emulsion Sprays on Leaf Surfaces," *Chem. Ind.* Nov. 10, 1962, p. 1917.
3. EATON, *op. cit.* p. 1915.
4. FARMIDGE, *op. cit.* p. 1917.
5. "New Pesticide Spray Methods Due This Spring," *Chem. Engr. News*, March 28, 1966, p. 42.
6. Wayland Chemical Corp.
7. Rohm and Haas Co.
8. The Dow Chemical Co.
9. Rohm and Haas Co.
10. The Dow Chemical Co.
11. Wayland Chemical Corp.
12. The Dow Chemical Co.
13. H. BENNETT, *Chemical Formulary, vol.* 11, p. 155, Chemical Publishing, New York.
14. The Dow Chemical Co.
15. Rohm and Haas Co.
16. BENNETT, *op. cit.* vol. 11, p. 155.
17. The Dow Chemical Co.
18. Rohm and Haas Co.
19. Mona Industries, Inc.
20. US Patent 2 153 487.

21. MacRill, US Patent 2 019 758 (1935)
22. Trout, *J. Coun. Sci. Ind. Res.* 15: 26 (1942).
23. Atlas Chemical Industries, Inc.

chapter 2

BITUMINOUS EMULSIONS

Bituminous emulsions are generally used as surface coatings. These emulsions are used in addition to fibers for protection of roads and highways. They are characterized by a low concentration of a relatively inexpensive surfactant.

Cationic surfactants are used for bituminous emulsions. The cationic material is strongly adsorbed at negatively-charged mineral surfaces. The mineral surface becomes water repellent because of the adsorbed cationic surfactant. The bitumen will adhere better to this water repellent surface and therefore can more effectively fulfill its purpose of holding together the mineral particles in the road surface. Thus, the cationic surfactants are often referred to as adhesion promoters. In use, the cationic bitumen-in-water emulsion is applied to the stone road surface. It breaks upon contact with the road surface and the bitumen follows the cationic surfactant to adhere to the stone surface, while the water runs away.[22]

Various anionic surfactants are also used for surface dressing of roads. Based upon the bitumen, 0.5 to 2.5% surfactant is added to asphalt and tar macadam basecourses and fine cold asphalt to protect the binder-to-stone bond from water action. Britain and Sweden make use of this technique to prevent stripping of oiled gravel surfaces, in addition to the above road surfaces.[23]

Asphalt Emulsions

Formula No. 1[1]

Asphalt	64.05
Water	35.00
Caustic Soda	0.08
Corn Gluten (or Soya Bean Meal)	0.64
Green Acid Soap (Dry Basis)	0.23

The asphalt for use in this formula may be produced from Mid-Continent petroleum, melting point about 110°F, penetration about 130 at 77°F.

Green acid soap may be prepared by the neutralization of green acids, which are well known in the petroleum industry. If the green acid soap contains appreciable amounts of oil, a harder asphalt should be emulsified to produce a residue of given penetration.

The caustic, farinacious emulsifier, and green acid soap are mixed with water and heated to a temperature of about 200°F. This hot solution or mixture is placed in a suitable stirrer, agitator, or mixer and is beaten by paddles, circulated by centrifugal pumps, or dispersed between suitable rotors moving at high velocities.
The melted asphalt at about the same temperature is slowly stirred into the solution and further agitated until complete emulsification has taken place. When high melting point asphalts are used, it may be necessary to increase the temperature at which the asphalt is added, but it is desirable to keep the temperature of the emulsion below the boiling point of water so that the foaming due to the production of steam may be prevented.

No. 2[2]

A sodium oleate solution is made up to a concentration of 20% by the addition of oleic acid and caustic soda to water at 90°C. This is then diluted with nine times its volume of water heated to 90°C. The 2% soap solution is run through a colloid mill with an equal amount of asphalt heated to a temperature not exceeding 100°C. The resultant emulsion contains equal parts of asphalt and water with 1% by weight of soap.

No. 3[3]

A hot dilute aqueous solution of alkali, such as a solution of caustic soda of about 0.5% strength, is prepared. An approximately equal weight of asphalt is melted; part of the melted asphalt is slowly stirred into the hot solution until scum begins to form on the surface; then a small quantity (about 0.5% of the final product) of oleic acid is added, followed by addition of the rest of the asphalt, while the temperature and agitation are maintained and a small proportion of clay added to give desired stability and adhesiveness.

No. 4[4]

A.	Rosin	100	C
	Slaked Lime	$3\frac{3}{4}$	
B.	Fuel Oil	103	

Heat A at 140°C and mix until uniform. Add B while mixing. Take 4 lb of C and add to 100 lb of melted bitumen or asphalt. Disperse the mixture in 0.05 N potassium rosinate solution to give a 57% bitumen content.

No. 5[5]

Asphalt	48-52
Water	46-48
Oleic Acid	1
Sodium Hydroxide	<1
Bentonite	1

No. 6[6]

Nine parts of soap are dissolved in 78 parts of warm water. About 20 parts of a low-grade fuel oil or a crude oil with an asphaltic base are added slowly, with vigorous agitation.

No. 7[7]

Asphalt	500
Water	500
Bentonite	30
Quebracho	30
Soda Ash	10

The bentonite, quebracho, soda ash, and water are mixed and heated to 200°F. While stirring, the asphalt, which has been heated to approximately 200°F, is added. The stirring is continued until the asphalt is dispersed.

No. 8[8]

Asphalt	2800
Water	2800
Rosin Soap (50%)	118
Pine Oil	40

The rosin soap is added to the water and is heated to 200°F. The asphalt is also heated to 200°F and the pine oil added. While agitating, the asphalt is slowly poured into the water. The mixture is agitated until a smooth emulsion is formed.

No. 9[9]

Hydrous Magnesium Silicate	8-10 lb
Water	20 gal

Mix well and heat to boiling; then mix in an emulsifying machine with

Asphalt, melted	36 gal
Acetic Acid (0.1 N)	250-240 cc
Water, boiling	10 gal

No. 10[10]

Adhesive for Sound-Deadening Pads

Asphalt	50-60
Bentonite	2-3
Oxalic Acid	0.02
Kerosene	3-10
Water	To make 100

The asphalt is warmed in the kerosene until dissolved. The bentonite and oxalic acid are dissolved in water and heated to boiling. The solutions are vigorously mixed and run through a colloid mill, if necessary.

No. 11[11]

Millboard Adhesive

Asphalt	100
Tall Oil Soap (50% water)	30
Kaolin	30

The tall oil soap and kaolin are mixed while heating and the molten asphalt slowly added, with stirring.

Before use, hot water in the desired quantity is added.

No. 12[12]

Soft Asphalt Emulsion

Water	400
Carbopol 941	1.5 (0.19%)
Sodium hydroxide (10% solution)	4.5
Ethomeen C-25	0.75
Soft Asphalt*	400

* Sohio 180-200 asphalt, Standard Oil Co. of Ohio, Cleveland, Ohio.

Carefully disperse the Carbopol 941 in the water. After thorough dispersion, neutralize with the sodium hydroxide followed by addition of the Ethomeen C-25. Heat this mucilage to 65-70°C. (The water can be heated prior to the addition of the Carbopol 941 and neutralizer as this will assist dispersion of the Carbopol 941.) Separately heat the asphalt to 85°C. Utilizing moderately high shear mixing, slowly add the molten asphalt to the water mucilage. Add the asphalt no faster than the rate of dispersion. If the asphalt is added too rapidly, the emulsion will invert and separate.

When the asphalt has been added, continue mixing for a short time to insure uniformity. Shock cool the emulsion without mixing and allow it to remain undisturbed until its temperature falls to 30-35°C.

	No. 13[13]	No.14[13]
A. Asphalt†	65	65
B. Water	35	35
C. Diam 26	0.1-0.2	-----
Diam 11-C	--------	0.25-0.30
D. Conc. Hydrochloric Acid††	0.05-0.15	0.15-0.20

For optimum stability and minimum processing costs, the emulsion should be formed inside a colloid mill with the water and asphalt fed to the mill separately. Two methods of accomplishing this follow. The first method leads to somewhat better emulsions but requires HCl-resistant processing equipment. The amine salt used in Method II is noncorrosive.

METHOD I

(1) Dilute D in B

(2) Add C to A. The asphalt and water should be in separate tanks and independently connected to the mill through pumps.

† 150-200 or 200-250 penetration paving grade asphalt, 60 to 70 parts of asphalt, may be used provided the emulsifying agent concentration is adjusted to correspond.

†† pH is usually adjusted to 6.0-6.5.

(3) Simultaneously pump AC and BD into the colloid mill, at a proportion of 65/35.

METHOD II

(1) Melt C and add a portion of the water; finally, add D.
(2) Recycle with the pump to emulsify CD.
(3) Pump BCD into another tank containing the remainder of the water, agitate or recirculate.
(4) Heat BCD to 80-90°C and the asphalt to 115-120°C.
(5) Pump the asphalt and BCD into the colloid mill in proportions of 65/35.

No. 15[14]

1000 g of bitumen (preferably asphalt) are heated to 90°C. 60 g of Swedish resin at 80-90°C are then added, and finally, a solution of 20 g of caustic potash in 120 g of water, cooled to 60-70°C. Vigorous stirring is needed throughout. 1000 g of hot water are poured in and when soap formation is ended, 800 g of water are added.

No. 16[15]

Add 15 parts by weight of oleic acid to 250 parts by weight of asphalt flux oil, heating the mix to about 100°C. Add this to 750 parts of water to which have been added 34 parts of aqueous ammonia, to form an emulsion of the asphalt flux oil, which has a viscosity substantially greater than that of water. Add 1500 parts by weight of coal tar, specific gravity of about 1.18 or more, heated to a temperature of about 70°C, to which has been added 45 parts of oleic acid. Vigorously agitate the tar with the emulsion of asphalt flux oil and subject the resulting mixed emulsion to intensive mechanical disintegration —for example, by passage through a colloid mill.

No. 17[16]

Emulsion for road making

Spramex Bitumen	48.0
Water	49.5
Sodium Carbonate (calcined)	0.5
Oleic Acid	2.0

The bitumen is warmed at 95-98°C and the oleic acid added. The water is heated separately with the sodium carbonate and the two liquids are introduced into the emulsifier.

No. 18[16]

Spramex Bitumen	50.0
Mineral Oil	2-2.5
Resin Soap	1.5-2.0
Caustic Potash	1.0
Water	45.0

The bitumen is melted and the mineral oil added during agitation. The water is heated to boiling and in it are dissolved the soap and the caustic potash. The liquids are mixed at 95°C.

With more bitumen, 1-2% of blue starch, gelatin, or sodium silicate must be added during or after emulsification.

No. 19[17]

Melt 100 parts of bitumen, softening point 45-50°C, and add 10.8 parts of rosin; heat to 100-125°C. Then add 20 parts of kaolin (soaked in equal parts by weight with water) and 1.2 parts of sodium hydroxide, preheated to 70-80°C. Raise the temperature of this mixture to 100°C and dilute with water to produce an emulsion containing 20-25% solids.

No. 20[18]

Bitumen	35
Coal, powdered	15
Carragheen	1
Water	50

No. 21[19]

Pitch Emulsion

A.	Coal Tar Pitch	180
	Stearin Pitch	20
B.	Casein Solutlon*	45
	Water	35
	Caustic Potash	1
C.	Water, boiling	120

Melt A together and add to B at 100°C. Add C.

* *Casein Solution*

Casein	56
Caustic Potash (50%)	10
Water	494
Cresylic Acid	11

Tar Asphalt Emulsion

No. 22[20]

Shale Tar	38.15
Water	49.60
Mexican Asphalt	7.63
Casein	1.15
Rosin	1.15
Potato Starch	0.76
Anthracene Oil	1.34

Tar Emulsions

No. 23[21]

Water	50.000
RT-12 tar	50.000
Carbopol 941	0.125
A. Sodium hydroxide (10% solution)	0.375
Ethomeen C-25	0.625

Carefully disperse the Carbopol 941 in the water in a jacketed vessel, and add A. Heat both the water mucilage and the tar to 65°C. Slowly add the molten tar to the hot mucilage with vigorous mixing (a mixer such as a medium-speed Eppenbach homogenizer is recommended). After the last of the tar has been added, mix briefly to ensure homogeneity. Rapidly cool the emulsion without further mixing.

The mixing step is critical. With very high shear mixing, the tar particles are extremely small and the emulsions are very smooth. Films deposited from such emulsions are difficult to dry, however, because an impermeable layer forms on the top. In contrast, emulsions formed with moderate shear have relatively large tar particles but dry evenly and rapidly.

No. 24[21]

Water	50.000
RT-8 tar	50.000
Carbopol 941	0.125
Sodium hydroxide (10% solution)	0.250
Ethomeen C-25	0.063

Carefully disperse the Carbopol 941 in the water. When dispersion is complete, add the sodium hydroxide and then the amine. Lastly, slowly add the RT-8 tar in a thin, continuous stream (so that the rate of addition is no greater than the rate of dispersion) with good mixing. Fairly high shear, such as that provided by a medium-speed Eppenbach homogenizer, forms the best (smallest droplet-size) emulsion. Excessively rapid addition of the tar results in a grainy emulsion.

No. 25[21]

Creosote Emulsion

Water	200.00
Carbopol 934	0.60
Sodium Hydroxide (10% solution)	0.55
Creosote	200.00

Carefully disperse the Carpopol in the water and add the sodium hydroxide. Add the creosote in a slow stream while vigorously agitating the mix. Stir to uniformity.

References

1. H. BENNETT, *Chemical Formulary*, vol. 2, p. 187, Chemical Publishing, New York, 1935.
2. H. BENNETT, *Ibid.* vol. 1, p. 163, Chemical Publishing, New York, 1933.
3. Personal Communication.
4. BENNETT, *op. cit.* vol. 4, p. 97, Chemical Publishing, New York, 1939.
5. *Ibid.* vol. 2, p. 186.
6. *Ibid.* p. 186.
7. *Ibid.* vol. 1, p. 155.
8. *Ibid.* p. 155.
9. *Ibid.* vol. 4, p. 96.
10. US Patent 2,333,779.
11. BENNETT, *op. cit.* vol. 7, p. 39 Chemical Publishing, New York, 1945.
12. B. F. Goodrich Chemical Co.
13. General Mills.
14. BENNETT, *op. cit.* vol. 4, p. 39 Chemical Publishing, New York, 1945.
15. *Ibid.* p. 97.
16. *Ibid.* vol. 1, p. 329.

17. *Ibid.* vol. 5, p. 72.
18. *Ibid.* p. 72.
19. *Ibid.* vol. 4, p. 97.
20. *Ibid.* vol. 5, p. 73.
21. B. F. Goodrich Chemical Co.
22. D. H. MATHEWS, "Surface Active Agents in Bituminous Road Materials," *J. App. Chem.* 12:56-64 (1962).
23. *Ibid.*

chapter 3

CLEANERS

Light Duty Liquids

Formula No. 1[1]

Dodecylbenzene Sulfonate, Sodium Salt	17
Dodecylbenzene Sulfonate, Triethanolamine Salt	6
Nonylphenol-Ethylene Oxide Sulfate, Ammonium Salt	5
Dowfax 9N15	7
Water (dye and perfume)	65

No. 2[1]

Dowfax 9N9	8
Lauric Diethanolamide	2
Dodecylbenzene Sulfonate, Sodium Salt	15
Ethanol	5
Water (dye and perfume)	70

No. 3[2]

Liquid Heavy Duty All-Purpose Cleaner

The following cleaner is useful for greasy machinery, whitewall tires, upholstery and rug spot removal, greasy pots and pans, etc.

Alkyl Aryl Sulfonate (Ultrawet 35K)	31
Deriphat 154	6
Alkanolamide (Ninol AA62 Extra)	4
Tetrapotassium Pyrophosphate	20
Water	39

No. 4[3]

All-Purpose Liquid Detergent

Conco AAS-35	10.0
Conco PXS or SXS	10.0
Condensate PE	3.0
Tetrapotassium Pyrophosphate	20.0
Sodium Metasilicate	2.0
Water	55.0

All Purpose Cleaners

No. 5[4]

A.	Tetrapotassium Pyrophosphate	5.0
	Tetrasodium Pyrophosphate	5.0
	Water	66.0
B.	Potassium Hydroxide 45% Liquid	4.0
C.	Tall Oil Fatty Acid	10.0
D.	Miranol C2M-SF conc.	10.0

Heat A at 60-70°C until completely dissolved and add B. With agitation and heat, add C and when C is completely dispersed, add D. Mix until uniform and allow to stand until clear.

No. 6[5]

Household

Tetrapotassium Pyrophosphate	8
Dodecylbenzene Sulfonate, Triethanolamine Salt	4
Dowfax 9N9 Surfactant	4
Lauric Diethanolamide	2
Sodium Xylene Sulfonate	6
Water	76

No. 7[5]

Industrial

Tetrapotassium Pyrophosphate	4
Dodecylbenzene Sulfonate, Triethanolamine Salt	4
Dowfax 9N15 Surfactant	5
Lauric Diethanolamide	1
Water	86

No. 8[6]

Ammoniated

A. P and G Amide No. 72	5.0
Coconut Oil Fatty Acids	7.1
B. Potassium Toluene Sulfonate	5.0
Tetrapotassium Pyrophosphate	6.7
Lytron 602 (*Monsanto*)	0.2
Ammonium Hydroxide (28°Bé)	1.0
Water	80.4

Mix A. Add two-thirds of the water to th eammonium hydroxide, mix with A and heat to 100°F. To this solution add B and stir until completely dissolved. Take the remainder of the water, dissolve the Lytron 602, and add to the above with stirring until a uniform product is obtained.

No. 9[7]

Household Ammonia (Opacified)

26°Bé Aqua Ammonia	5
A. Dowfax 9N9 Surfactant	1
Dowfax 9N4 Surfactant	1
Water	93

Mix A and add to water with agitation; add ammonia.

No. 10[8]

Pine Oil Base Cleaner

Olate Flakes	10.0
Water	10.0
Pine Oil	80.0

Mix the Olate Flakes in cool, soft water and warm to speed the rate of solution. Add the pine oil and mix thoroughly.

No. 11[8]

Pine Oil Disinfectant Cleaner

A. Olate Flakes	7.0
P and G Amide No. 23	3.0
Dowicide A	4.0
B. Pine Oil	10.0
Isopropyl Alcohol (99%)	10.0
Water	66.0

Dissolve A in water and agitate while heating to about 140°F. Add the Dowicide A, followed by B. Mix well until a clear solution is obtained.

DISHWASHING LIQUIDS

Hand Dishwashing Detergents

No. 12[9]

Benax 2A1	10.0
ABS	20.0
Alipal CO-436	5.0
Lauric Diethanolamide	3.0
Water	62.0
Dye, Perfume	*q.s.*

No. 13[10]

A. Orvus K Liquid	36.0
Kyro EOB	4.0
Lytron 624	0.5
Water	59.5

Mix A and add three-fourths of the water. Stir until completely dissolved. Disperse the Lytron 624 in the remaining one-fourth of the water and mix the two solutions until thoroughly dispersed. No heating is necessary.

No. 14[11]

A. Concopal A	20.00
Conco AAS-40	50.00
Condensate PE	6.75
Conco SXS	10.00
B. Morton E-159	2.80
Water	8.45

Mix A and add B.

No. 15[12]

Miranol CS conc.	15.0
Tridecyl Triethoxy Sulfate (60% active)	10.0
Lauric Acid Dialkanolamide 90%	2.5
Water	72.5

Machine Dishwashing

No. 16[13]

Plurafac RA-20 or RA-30	1-3
Sodium Tripolyphosphate	30-35
Soda Ash	30-35
Sodium Metasilicate (anhydrous)	15-25
Sodium sulfate	10-20

MISCELLANEOUS KITCHEN CLEANERS

Oven Cleaners

No. 17[14]

Deriphat 151	5
Methocel (thickener)	4
Water	91

No. 18[15]

Caustic Soda	10.0
Benax 2A1	0.5
Starch	7.0
Water	82.5

No. 19[16]

Rinse Aid

The rinse aid is either used "as is" or diluted with water or ethyl alcohol, depending upon the type of injection device employed. A diluted product is not necessary when it is used with an injector equipped with a dilution tank.

Plurafac RA-20 or RA-30	50-100
Ethyl Alcohol	0-50
Water	0-50
Quaternary Germicide	0-0.3
Color	0-trace

HOUSEHOLD DETERGENTS

Light-Duty Detergents

No. 20, 21, 22

	20[16]	21[16]	22[16]
Plurafac A-16, B-16, or C-17	6-10	6-10	6-10
Linear Alkylate Sulfonate (60% active)	20-25	20-25	----
Hydrotrope (Sodium Toluene Sulfonate or Sodium Xylene Sulfonate) (100% active)	----	5-10	5-10
Alcohol Ether Sulfate (50% active)	12-15	6-10	20-35
Fatty Amide	2-4	----	----
Isoproqyl Alcohol	5-9	----	----
Opacifier	2-5	----	----
Water	35-45	45-60	45-60

No. 23, 24

Granular Detergent

	23[16]	24[16]
Plurafac C-17 or B-16	6-12	6-12
Linear Alkylate Sulfonate (40% active flake)	35-50	40-60
Alcohol Ether Sulfate (60% active)	10-20	-----
Fatty amide	1-2	1-2
Sodium Sulfate and/or Sodium Bicarbonate	15-35	20-40

Spray liquid ingredient on dry ingredient in a horizontal ribbon mixer.

Heavy-Duty Detergents

No. 25[16]

Plurafac B-26 or D-25	8-15
Carbose D (65% NaCMC)	1-2
Sodium Tripolyphosphate	30-40
Sodium Metasilcate (anhydrous)	5-10
Soda Ash and/or Sodium Sulfate	33-50

No. 26[16]

Plurafac A-88	8-15
Plurafac B-26 (optional deduster)	0-1
Sodium Tripolyphosphate	30-40
Sodium Metasilicate (anhydrous)	5-10
Carbose D (65% NaCMC)	1-2
Soda Ash and/or Sodium Sulfate	32-50

No. 27[16]

Heavy Duty Detergent Tablet

Plurafac A-38 or A-28	8-15
Sodium Tripolyphosphate	50-65
Sodium Metasilicate (anhydrous)	5-10
Carbose D (65% NaCMC)	1-2
Soda Ash and/or Sodium Sulfate	10-30

No. 28[16]

A.	Water	to 100
B.	Plurafac B-16, B-26, or D-25	8.0-10.0
C.	Gantree AN-149	0.5-0.6
D.	Potassium Hydroxide (50% solution)	3.6-4.0
E.	Carbose D (65% NaCMC)	0.5-0.6
F.	Sodium Silicate (Water Glass)	5.0-10.0
G.	Tetrapotassium Pyrophosphate (60% solution)	32.0-42.0

Mix A, 0.01% of B, and C in that order with gentle agitation. Heat ABC to 80°C (176°F) and maintain at that temperature for 20-30 minutes or until the vicous mixture becomes fluid. With gentle agitation add D and when the mixture has become homogenous, add E. After the Carbose has been dissolved, add F. Cool the solution to 60°C (140°F) and with vigorous agitation rapidly add the balance of B. Continuing the vigorous agitation, add G (preheated to 60°C) very rapidly. If desired, optical brightener, perfume, and dye are added immediately and the vigorous agitation continued for another five minutes. The final product can then be cooled and withdrawn into suitable containers.

No. 29[17]

Miranol OS	40.0
Sodium Xylene Sulfonate	2.0
Tetrapotassium Pyrophosphate	20.0
Potassium Hydroxide-45% Liquid	5.0
Propylene Glycol	3.0
CMC-7AA (*Hercules Powder Co.*)	0.5
Kessco X435 (*Kessler Chemical Co.*)	2.0
Cab-O-Sil M-5 (*Cabot Corp.*)	2.0
Kasil No. 1 (*Philadelphia Quartz Co.*)	10.0
Lauric Acid Dialkanolamide 90%	2.0
Water	13.5

No. 30[17]

Household Laundry Compound

Miranol CS conc.	25.0
Kasil No. 1 (*Philadelphia Quartz Co.*)	10.0
Tetrapotassium Pyrophosphate	20.0
CMC-7AA (*Hercules Powder Co.*)	0.5
Propylene Glycol	3.0
Potassium Hydroxide (45% liquid)	4.5
Cab-O-Sil M-5 (*Cabot Corp.*)	2.0
Kessco X435 (*Kessler Chemical Co.*)	2.0

Lauric Acid Dialkanolamide (90%)	2.0
Water	31.0

Auxiliary and Special Purpose Laundering

No. 31[18]

Consumer Laundry Softener

Conco Softener CC-75	9.0
Tinopal BHS	0.2
Perfume	*q.s.*
Coloring	*q.s.*
Water	to 100

Fine Fabric Detergents

No. 32[19]

Dowfax 9N9	10
Lauric Diethanolamide	1
Dodecylbenzene Sulfonate, Sodium Salt	4
Dodecylbenzene Sulfonate, Triethanolamine Salt	4
Sodium Tripolyphosphate	4
Benax TM 2A1 Surfactant	2
Water (dye, perfume)	75

No. 33[19]

Benax 2A1	18.0
Tetrapotassium Pyrophosphate	20.0
Lauric Diethanolamide	5.0
Optical Brighteners, Dye, Perfume, Water	to 100.0

No. 34[20]

Liquid Woolen Cleaner

Conco AAS-70	24.0
Conco Sulfate TL	5.0
Urea	2.0
Ethanol	3.0
Water	45.85
Tinopal RBS-200	0.05
2% Methocel Solution (HD-62)	20.0
Perfume	0.1

Hydrogen Peroxide Emulsions

No. 35-39

The following formulas range from thin to heavy lotions:

	35[21]	36[21]	37[21]	38[21]	39[21]
Arlacel 83	5.7	11.4	12.0	18.0	17.0
Tween 40	----	----	8.0	12.0	----
Tween 80	4.3	8.6	----	----	13.0
Hydrogen peroxide (35%)	23.0	23.0	23.0	23.0	23.0
Water	57.0-67.0	57.0	57.0	47.0	47.0
Stabilizer	(Type and quantity as required by formulator)				

Mix surfactants, add H_2O_2, and blend. Add water and blend. **WARNING**: Extreme care is essential in formulating with hydrogen peroxide because rapid release of oxygen can occur under heat or contamination, resulting in explosion of containers. Any composition should be adequately stabilized to avoid danger.

No. 40[22]

A. Deionized water	863.9
B. Acetophenetidin, USP	0.4
C. Cetyl alcohol	25.0
Arlacel 165	25.0
Hydrogen peroxide (35%)	85.7

Add the acetophenetidin to 400

parts of boiling deionized water (A). Mix C and heat to 70°C, agitate C while adding 300 parts of deionized water that has been previously heated to 72°C. Add B to A with mechanical agitation, using part of the remaining water to make sure all of B is rinsed into A; allow the mixture to cool to room temperature. Add the remaining water and if necessary compensate for water lost in boiling. Add hydrogen peroxide and adjust the pH to 3.5-4.0 with phosphoric acid (10%)

FLOOR AND RUG CARE PRODUCTS

Floor Wax Removers

No. 41[23]

A.	Oleic Acid	11.0
	Dowfax 9N9 Surfactant	10.0
B.	Monoethanolamine	3.0
C.	Versene® 100 Chelating Agent	1.5
	Trisodium Phosphate	8.0
	26°Bé Ammonia	5.0
Water		61.5

Mix A and add B, stirring to form a clear system. Add C to the water. Add the solution of C and water to AB with agitation.

No. 42[23]

Acid Wax Stripper

Dowanol EB Glycol Ether	50
Phosphoric Acid (85%)	25
Dowfax 9N10 Surfactant	25

No. 43[24]

Acid Wax Stripper

Miranol C2M-SF conc.	2.0
Hydroxy Acetic Acid	3.6
Cellosolve	4.0
Water	90.4

No. 44[25]

Basic Wax Stripper

Benax 2A1	6.0
Tetrapotassium Pyrophosphate	4.0
Sodium Metasilicate	4.0
Versene 100	0.5
Water	85.5

No. 45[26]

Sanitizing Floor Cleaner

Monateric C-Na (40%)	10
Dowicide 32	10
Diethanolamine	2
Sodium Hydroxide Flakes	2
Tetrapotassium Pyrophosphate	6
Water	70

Rug Cleaners

No. 46[27]

Moderate-to-Low Foam

Benax 2A1	6.8
Dowanol EB	3.2
Potassium Oleate	1.0
Water	89.0

No. 47[27]

High Foam Rug Cleaner

Benax 2A1	6.8
Dowanol® EB Glycol Ether	3.2
Lauric diethanolamide	1.0
Water	89.0

No. 48[28]

A.	Oleic Acid	35
	Trigamine	15
	Water	125
B.	Butyl Cellosolve	5
	Ethylene Dichloride	13
C.	Diethylene Glycol	15
	Isopropyl Alcohol	20

Mix A and B with a high-speed mixer. Add C slowly. This gives a clear solution which is readily emulsifiable in water.

No. 49[28]

A.	Oleic Acid	28
	Ethylene Dichloride	13
	99% Isopropanol	14
	Butyl Cellosolve	5
B.	Triethanolamine	16
Water		125

Mix A and add B. Stir AB until thoroughly mixed and add the water. If the mixture is cloudy, add sufficient isopropanol to clear it.

Rug Shampoos

No. 50[29]*

Plurafac C-17 or B-16	2-5
Sodium Tripolyphosphate	1-2
Isopropyl alcohol	5-10
Water	85-90

No. 51[29]*

Dry-Powder Rug Shampoo

After the shampoo solution dries on the fabric, the dry powder residue can be readily vacuumed.

Plurafac A-38	20-40
Sodium Sulfate	30-40
Sodium Sesquicarbonate	40-60

* Dilute these formulations with water 1 to 10 for use on carpets and rugs or 1 to 20 for use on upholstery.

MISCELLANEOUS HOUSEHOLD CLEANERS

Bowl Cleaners

No. 52[30]

20°Bé HCl	73
Zinc Chloride	2
Conco NI-100	10
Conco Product 150	1
Water	13
Hyamine 1622	1

No. 53[31]

Bowl Cleaner-Sanitizer

Benax 2A1	10.0
Phosphoric Acid (85%)	11.8
Dowicide 32	1.5
Water	76.7

No. 54[31]

Glass Cleaner

Dowanol® DPM Glycol Ether	3.0
Dowanol PM Glycol Ether	8.0
Isopropanol	5.0
Dowfax® 9N9 Surfactant	0.5
Water (dye)	83.5

Shoe Cleaners

No. 55[32]

Naphtha Emulsion

A.	High-Viscosity Methyl Cellulose	1 g
B.	Water	100 ml
C.	2-Amino-2-Methyl-1, 3-Propanediol	2.3 g
D.	Stearic Acid	6.2 g
E.	Low-boiling Naphtha	200 ml

Disperse A in B and add C. Dissolve D in E. Pour DE into ABC and agitate. Better mixing can of course be obtained with a homogenizer, particularly in larger batches.

No. 56[33]

Shoe White

Condensate PS	3.0
Titanium Dioxide	36.0
Water	61.0
Gum Arabic	1.0-1.5

No. 57[34]

Leather and Plastic Cleaner

A.	Silicone Oil*	5.0
	Oleic Acid	2.0
B.	Mineral spirits	20.0
C.	Morpholine	1.0
	Renex 20†	2.0
Water		69.8
Carbopol 934		0.2

Dissolve A in the mineral spirits. C is dissolved in 50 parts of the water with mild agitation. Carefully disperse the Carbopol 934 in the remainder of the water. Add AB to the C solution with good agitation. Lastly, add the Carbopol 934 dispersion and stir to uniformity. Apply the cleaner with a clean cloth. Follow this by buffing.

* L-45 (350cS) *Union Carbide Chemicals Corp., Silicone Div.*, 270 Park Avenue, New York, N.Y. 10017.

† *Atlas Chemical Industries, Inc.*, Wilmington, Del. 19899.

No. 58[35]

Lotion Type Pet Shampoo

Orvus K Liquid	40.0
Ethylene Glycol Distearate	2.0
Lanogel 31	2.0
Water	56.0

Mix all of the ingredients and heat to 150°F in a covered vessel. Agitate continously and continue stirring until the material has cooled to 100°F. If a higher viscosity is desired, 1% ammonium chloride can be added.

Waterless Hand Cleaners

No. 59[36]

A.	Amerchol L-101	5.0
	Acetulan	3.0
	Polyethylene Glycol 600 Monostearate	3.0
	Polyethylene Glycol 400 Distearate	3.0
	Stearic Acid, XXX	4.0
	Mineral Oil, 70 vis.	10.0
	Odorless Mineral Spirits or Kerosene	35.0
B.	Water	30.0
	Triethanolamine	2.0
	Propylene Glycol	3.0
Zinc Stearate, USP		2.0
Perfume and Preservative		*q.s.*

Heat A to 65°C The mineral spirits should be incorporated in to A toward the end of this heating cycle to prevent excessive volatilization. Heat B separately to 65°C and add to A with mechanical agitation Add the zinc stearate at 55°C Cool with stirring to 45°C, perfume, and package.

Note: This formula has a set point below 42°C. The set point can be lowered by decreasing the amounts of polyethlene glycol esters and/or the stearic acid. The set point can be raised by increasing the concentration of the polyethylene glycol esters and/or by substituting PEG 600 distearate for the PEG 400 distearate.

No. 60[37]

A.	Mineral oil	39.0
	Oleic acid	9.0
B.	Dowfax 9N9	5.0
	Propylene Glycol	4.5
	Triethanolamine	2.5
	Monoethanolamine	1.0
	Water	39.0

Mix A. Mix B and add B to A slowly with constant stirring.

No. 61[38]

A.	Amber Granules®	15.0
	Kerosene or High Flash Naphtha	35.0
	Water	45.0
Cyclohexanol		5.0

Mix A and agitate while heating; then add the cyclohexanol.

No. 62[38]

A.	P and G Amide No. 72	10.0
	Kyro EOB	12.0
Light Mineral Oil		48.0
Water		30.0

Dissolve A in the water and stir in the mineral oil until a stable emulsion is formed. The emulsion will remain liquid and can be poured; however, upon standing, it firms up into a creamy white gel paste. **Note:** No heating is required.

No. 63,64

	63[39]	65[39]
Amerchol L-101	1.0	0.6
Deodorized Kerosene	42.5	43.0
Oleic Acid, Emersol 221*	6.0	6.1
Cetyl Alcohol	0.5	----
Triethanolamine	3.00	3.0
Propylene Glycol	2.35	3.2
Tergitol 4†	1.40	1.8
Perfume	0.25	0.3
Distilled Water	43.00	42.0

Aerosol as follows:

Above concentrate	85% by wt.
Genetron 12††	15% by wt.

* *Emery Industries, Inc.*
† *Union Carbide Chemicals Co.*
†† *Gen. Chem. Div., Allied Chem. Corp.*

Car Care Products

No. 65[40]

Monamine ALX-80%	1.0
Water	6.0

Amount to be used depends on working conditions—for instance, machine or handwashing, size of tanks, feed systems, etc. This product is free rinsing and nonsteaming.

No. 66[40]
(with Silicone)

Monamine ALX-80%	50.0
Water	47.5
G. E. Silicone SF 96 (*General Electric Co.*)	2.5

Mix the Silicone into the Monamine and add water last. A practically clear solution that will show only a negligible separation upon standing is obtained. This product should not be packed in unlined tin cans. Dilute for use, 1 or 2 ounces in 3 gallons of water.

No. 67[41]
(Powder)

Conco AAS-90P	30.0
Trisodium Phosphate	65.0
Sodium Chloride	5.0

No. 68[42]

Benax 2A1	14.0
ABS	12.0
Disodium Phosphate	12.0
Water	62.0

No. 69[42]
Windshield Washer Concentrate

Dowfax 9N9	1
Isopropyl alcohol	99

(Water is optional)

No. 70[42]
White Wall Tire Cleaner

Tetrapotassium Pyrophosphate	2.0
Sodium Orthosilicate	0.1
Dowanol PM Glycol Ether	5.0
Dowfax 9N10 Surfactant	1.5
Benax 2A1 Surfactant	1.5
Water	89.9

No. 71, 72
Engine Cleaners

		71[43]	72[43]
A.	Kerosene (Shell Insecticide Base)*	80.0	80.0
	Butyl Cellosolve	10.0	10.0
B.	Kyro EOB	5.0	-----
	P and G Amide No. 72	5.0	10.0

Blend A and add B. Stir until a clear solution is obtained.

* *Shell Chemical Co.*

Detergent Sanitizers

The combined characteristics of good detergency, moderate foam, and rinsability are important in detergent sanitizers.

No. 73[44]

Detergent Disinfectant

Benax 2A1	10.0
Tetrapotassium Pyrophosphate	10.0
Dowicide® 32 Germicide	5.0
Isopropyl Alcohol	3.0
Water	72.0
Phenol coefficient (Salmonella typhosa)	7.5
Phenol coefficient (Staphylococcus aureus)	4.7

No. 74[45]

Plurafac B-26 or D-25	8-10
Sodium Tripolyphosphate	20-30
Sodium Metasilicate (anhydrous)	10-20
Soda Ash	20-30
Quaternary Germicide	5-10

Note: An active chlorine-containing compound can, when properly formulated, replace the quaternary compound.

Metal Cleaners

Acid Aluminum Cleaners

No. 75[46]

Phosphoric Acid (85%)	15
Hydrofluoric Acid (48%)	2
Glacial Acetic Acid	3
Dowfax 9N10	5
Water	75

No. 76[46]

(can be used hot)

Dowanol EB	6.0
H_3PO_4 (85%)	7.0
Benax 2A1	0.1
Water	86.9

No. 77[47]

Alkaline Aluminum Cleaner

Miranol C2M-SF conc.	5-10
Sodium Metasilicate Pentahydrate	5
Water	85-90

No. 78[48]

Copper Cleaner

Dowfax 9N9	2
Sodium Chloride	5
Citric Acid	5
Bentonite	22
Water	66

No. 79[49]

Liquid Acid Equipment Cleaner

Miranol J2M conc.	1.8
Miranol C2M-SF conc.	0.2
Gluconic Acid	6.0
Phosphoric Acid 75%	62.0
Water	30.0

No. 80

Liquid Caustic Equipment Cleaner

Miranol J2M conc.	2.0
Water	47.5
Potassium Hydroxide 45% Liquid	0.5
Kasil® (*Philadelphia Quartz Co.*)	50.0

Degreaser Formulations

No. 81[50]

A.	Olate Flakes®	4
B.	Pine Oil (Yarmor)	10
	Water	35
C.	Kyro EOB	10
D.	Kerosene (Shell Insecticide Base)	26
E.	Butyl Cellosolve	15

Dissolve A in B. When completely dissolved, add C, D, and E in that order, stirring until the solution is clear. **Note:** Due to the presence of insoluble inhibitors in the Olate Flakes, a precipitate will form on standing. This can be removed, either by decanting off the upper clear portion or by filtering.

No. 82[51]

Kerosene	85.0
Conco AAS Special No. 3	15.0

No. 83[52]

Clear Emulsion Cleaner*

Trichlorethylene	100 cc
Triethanolamine Oleate	2 g
Monoethanolamine Oleate	6 g
Sodium Oleate	1 g
Water	11 cc

* U.S. Patent 2,576,419

No. 84[53]

Phosphoric acid 75%	30
Monazoline O	5
Kerosene	5
Water	60

The addition of oleic acid will greatly increase the corrosion inhibition of this formulation.

Steam Cleaners

No. 85[54]

Light Duty

Sodium Metasilicate	30.0
Sodium Tripolyphosphate	58.0
Trisodium Phosphate	10.0
Benax 2A1	1.0
Triton CF-10 Surfactant	1.0

No. 86[54]

Medium Duty

Sodium Orthosilicate	33.0
Sodium Carbonate	20.0
Trisodium Phosphate	44.0
Benax 2A1	1.5
Triton CF-10	1.5

No. 87[54]

Heavy Duty
(for oily soils)

Benax 2A1	7.0
Sodium Metasilicate	35.0
Dowanol DE	3.0
Water	55.0

Alkaline Cleaners

Alkaline-built cleaners are heavily relied upon to remove grease and oil soils.

No. 88[55]

Plurafac A-24, RA-20, or RA-30	1-2
Soda Ash	40-45
Trisodium Phosphate	10-15
Sodium Tripolyphosphate	20-25
Borax	15-20
Linear Alkylate Sulfonate	0-1

No. 89[55]

Plurafac B-16, B-26, or D-25	2-3
Sodium Metasilicate	45-55
Caustic Soda	5-10
Tetrasodium Pyrophosphate	5-7
Trisodium Phosphate	30-40

Note: An effective alkaline soak-type cleaner (used in concentrations of 4-8 oz/gal of water).

Miscellaneous Industrial Cleaning Formulations

Electrolytic Cleaning

"Electrolytic cleaning is often used when metal surfaces must be prepared for final finishing. If alkaline solutions are used, hydrogen gas is evolved at the cathode and oxygen at the anode, producing a scrubbing effect on the metal surface. Plurafac A-26 and A-28 (in the following formulation) serve as wetting agents in this process; being nonionic, they are unaffected by the current."[55]

No. 90[55]

Plurafac A-25 or A-28	1-2
Sodium Orthosilicate	40-55
Soda Ash	15-25
Caustic Soda	20-30

The bath concentration should be 6-18 oz/gal at a temperature range of 200-220°F, with a current density of 60-100 A/ft².

No. 91[56]

Food Peeling, Bottle Washing, etc.

Miranol C2M-SF Conc.	1.0
Carbitol® (*Union Carbide Chemicals Co.*)	1.0
Sodium Hydroxide Flakes	20.0
Water	78.0

No. 92[56]

Powdered Caustic Bottle Washing Compound

Miranol J2M conc.	1.0-2.0
Gluconic Acid	6.0
Caustic Soda Flakes	92.0-93.0

No. 93[57]

Mercerization Formulation

Benax 2A1	0.24
2-Ethyl Hexyl Alcohol	0.16
2-Methyl-2,4-Pentanediol	0.15
Sodium Hydroxide (20% aqueous solution)	99.45

No. 94[57]

Paper Machine Felt Cleaner

Concentrated HCl (36%)	10.0
Benax 2A1	0.3
Dowfax 9N9	0.2
Water	89.5

References

1. The Dow Chemical Co.
2. General Mills
3. Continental Chemical Co.
4. Miranol Chemical Co., Inc.
5. The Dow Chemical Co.
6. Proctor and Gamble, Co.
7. The Dow Chemical Co.
8. Proctor and Gamble, Co.
9. The Dow Chemical Co.
10. Proctor and Gamble, Co.

11. Continental Chemical Co.
12. Miranol Chemical Co., Inc.
13. Wyandotte Chemicals Corp.
14. General Mills
15. The Dow Chemical Co.
16. Wyandotte Chemicals Corp.
17. Miranol Chemical Co., Inc.
18. Continental Chemical Co.
19. The Dow Chemical Co.
20. Continental Chemical Co.
21. H. BENNETT, *Chemical Formulary*, vol. 12, p. 277, Chemical Publishing, New York.
22. Atlas Chemical Industries, Inc.
23. The Dow Chemical Co.
24. Miranol Chemical Co., Inc.
25. The Dow Chemical Co.
26. Mona Industries, Inc.
27. The Dow Chemical Co.
28. H. BENNETT, *op. cit.* vol. 11, p. 337.
29. Wyandotte Chemicals Corp.
30. Continental Chemical Co.
31. The Dow Chemical Co.
32. Commercial Solvents Corp.
33. Continental Chemical Co.
34. B. F. Goodrich Chemical Co.
35. Proctor and Gamble, Co.
36. American Cholesterol Products, Inc.
37. The Dow Chemical Co.
38. Proctor and Gamble Co.
39. American Cholesterol Products, Inc.
40. Mona Industries, Inc.
41. Continental Chemical Co.
42. The Dow Chemical Co.
43. Proctor and Gamble, Co.
44. The Dow Chemical Co.
45. Wyandotte Chemicals Corp.
46. The Dow Chemical Co.
47. Miranol Chemical Co., Inc.
48. The Dow Chemical Co.
49. Miranol Chemical Co., Inc.
50. Proctor and Gamble Co.
51. Continental Chemical Co.
52. H. BENNETT, *op. cit.* vol. 11, p. 331.
53. Mona Industries, Inc.
54. The Dow Chemical Co.
55. Wyandotte Chemicals Corp.
56. Miranol Chemical Co., Inc.
57. The Dow Chemical Co.

chapter 4

COSMETICS

Both oil-in-water and water-in-oil emulsions find wide use in the field of cosmetics. Foundation creams, cold creams vanishing creams, hand lotions and creams, sunscreen lotions and creams, hair creams, and shaving creams are a few of the uses of emulsified cosmetics. A small amount of an active ingredient is easily suspended in the emulsion form. Uniform application of this ingredient is achieved without discomfort or pain to the user. The emulsion form often increases the rate at which an active ingredient may penetrate the skin. The water-oil-surfactant blend of the emulsion can also be formulated for effective cleansing of both oil- and water-based contaminants. The cosmetic emulsion also allows the simultaneous application of water- and oil-soluble materials.

A cold cream is designed for cooling and cleansing the skin. Traditional cold creams are composed of beeswax, borax, and oil emulsified in 20-35% water. These creams are usually somewhat oily or greasy on the skin. Vanishing creams, hand creams, and shaving creams are examples of "greaseless creams." These creams are formulated to "disappear" when rubbed into the skin. Such oil-in-water emulsions must quickly rub into the skin without lathering, causing a clammy feel, or leaving an oily, tacky deposit. These products may contain significant amounts of skin softeners (e.g. glycerine), protective agents or lubricants (e.g. silicones), antiperspirants or deodorants, and perfumes.

Perfumes must be very carefully selected as they may react with the surfactant(s) and affect the stability of the emulsion, or oxidize and affect the color of the product. In addition, the odor balance of a perfume is changed when the perfume is incorporated into an emulsion. All ingredients in cosmetic preparations must be nonhazardous for their intended use or any likely misuse of the product. The Food, Drug and Cosmetic Act in the United States established the concern of the Food and Drug Administration for safe cosmetic products. The advisory section of the FDA is available to the emulsion formulator to aid him in the selection of permissible ingredients for cosmetic preparations.

Hand Lotions

No. 1[1]

A.	Amino-Methyl-Propanediol	0.8
B.	Stearic Acid	4.0
	Cetyl Alcohol	1.0
	Butyl Stearate	3.0
Quince Seed		0.5
Water		90.7
Preservative		as required

Heat A to 75°C with half the water. Heat B to 75°C and when both A and B are homogeneous, add B to A slowly with agitation. Soak the quince seed overnight with the rest of the water and the preservative; strain and stir the resulting "mucilage" into the cream (AB). The lotion thickens considerably after standing several days and may be thinned with water as desired Perfume is usually added in the last stages of manufacture as a highly concentrated solution in specially denatured alcohol.

No. 2[2]

A.	Water	862
	70% Sorbitol	41
	Ethanol (SD 40)	14
Carbopol 934		1
Butyl Parasept		1
Color		*q.s.*
B.	Triethanolamine	10
C.	Lanolin	17
	Cetyl Alcohol	3
	Stearic Acid	51

Completely disperse the Carbopol 934 in the blend A. When A is smooth and lump-free, add B. Heat both AB and C to 75°C and add C to AB with vigorous mixing. When the mix is uniformly blended, cool it rapidly to 30°C, adding perfume at 50°C.

No. 3[2]

A.	Water	835
	Propylene Glycol	83
	Propyl Parasept	1
Carbopol 934		1
Color		*q.s.*
Perfume		*q.s.*
B.	Triethanolamine	5
C.	Cetyl Alcohol	50
	Isopropyl Myristate	25
	Ethomeen C-25	1

Completely disperse the Carbopol 934 in A; add B Form C by melting the ingredients together. Heat AB and C to 75°C and add C to AB with vigorous mixing. After this has been blended to uniformity, cool it rapidly to 30°C. Add perfume at 50°C. The emulsion is ready for packaging or may be refined by homogenizing.

No. 4[3]

O/W Roll-Ball Hand Lotion

A.	Paraffin	22.5
	Stearic Acid	1.5
	Arlacel 40	1.4
	Tween 40	1.5
B.	Water	73.1
	Preservative	*q.s.*
C.	Perfume	*q.s.*

Melt A and bring to 70°C. Heat B to 75°C, adding to A a little at a time with moderate agitation. The first additions of water should be in extremely small portions with thorough agitation until the mixture inverts to O/W emulsified form. This is important for good stability. When all the water has been added, stir until cold and perfume.

No. 5[3]

O/W Hand Lotion

A.	Stearic Acid	7.0
	Lanolin	0.5
	Arlacel 80	0.5
	Tween 60	2.5
B.	Sorbo 70% Sorbitol Solution	10.0
	Preservative	*q.s.*
	Water	79.5
C.	Perfume	*q.s.*

Add A at 90°C to B at 95°C. Stir rapidly to 30°C. Perfume at 50°C and package.

No. 6[4]

Glyceryl Monostearate (self emulsifying)	2.7
Cetyl Alcohol	1.5
Silicone 200 (*Dow Corning Corp.*)	1.5
Lanolin Oil	2.0
Robane®	3.0
Sodium Lauryl Sulfate	0.3
Preservative	0.2
Water	to 100.0

No. 7[5]

Aerosol Hand Lotion

Modulan	1.5
Acetulan	1.0
Stearic Acid XXX	5.0
Cetyl Alcohol	1.0
Deltyl Extra	0.5
Triethanolamine	2.0
Propylene Glycol	4.0
PVP, Type NP-K30	0.3
Water	84.7
Perfume and preservative	*q.s.*

To Aerosol:

Above concentrate	92.0
Genetrons 114A/12 (43:57)	8.0

Hand Creams

No. 8[6]

A.	Glycerine	120.0
	Propyl Parasept	1.5
	Water	687.0
	Carbopol 934	2.0
	Color	*q.s.*
B.	Triethanolamine	1.7
C.	Cetyl Alcohol	100.0
	Lanolin	60.0
	Mineral Oil	20.0
	Ethomeen C-25	5.0

Carefully disperse the Carbopol 934 in A and add B. Heat AB and C to 75°C and add C to AB with vigorous mixing. When the blend has been mixed to uniformity, cool it quickly to 30°C. Add the desired perfume at 50°C.

No. 9[7]

O/W Hand Cream

A.	Stearic Acid	15.0
	Isopropyl Myristate-Palmitate	2.0
B.	Potassium Hydroxide	1.0
	Sorbo (70% sorbitol solution)	18.3
	Water	63.7
	Preservative	*q.s.*
Perfume		*q.s.*

Add B at 82°C to A at 80°C slowly under continuous agitation. Continue agitation to 45-50°C. Perfume and package.

No. 10[7]

O/W Hand Cream

A.	Stearic Acid	10.0
	Arlacel 60	5.2
	Arlacel 80	1.0
	Tween 60	2.8
B.	Water	81.0
	Preservative	*q.s.*
Perfume		*q.s.*

Add A at 80°C to B at 85°C with rapid stirring. Perfume at 50°C Pour at 45°C.

No. 11[7]

A.	Arlacel 165 acid-stable gms.	5.0
	Stearyl Alcohol	5.0
B.	Sorbo (7% sorbitol solution), USP	5.0
	Water	85.0
Mild acid to adjust pH		*q.s.*

Heat A to 70°C, B to 72°C. Add B to A with agitation. Stir until until cream has set.

No. 12[8]

Glyceryl Monostearate (self emulsifying)	4.0
Stearic Acid	4.0
Cetyl Alcohol	2.0
Lanolin	2.0
Robane®	4.0
Propylene Glycol	3.0
Triethanolamine	1.0
Preservative	0.2
Water and Perfume	to 100.0

Aerosol Hand Creams

No. 13[9]

A.	Polawax	2.0
	Solan Ethoxylated Lanolin	3.0
	Novol Oleyl Alcohol (deodorized)	3.0
	Stearic Acid (triple pressed)	2.5
B.	Propylene Glycol	5.0
	Triethanolamine	1.0
	Deionized Water	83.5
Above concentrate		92.0
Propellent 12/114 (60:40)		8.0

Heat A and B separately to 60-70°C and add B to A with stirring. Stir and cool down to about 40°C and fill aerosol containers with the concentrate. For maximum stability, add the propellents while the concentrate is still warm.

No. 14[10]

Acetulan	3.0
Solulan 75	3.0
Stearic Acid	2.6

Glyceryl Monostearate	3.0
Preservative	0.1
Triethanolamine	1.0
Propylene Glycol	3.0
Water	84.0
Perfume, Fleuralia No. 73*	0.3
Above concentrate	92.0
Propellents 114/12 (40:60)	8.0

* *Dodge and Olcott, Inc.*

Protective Hand Creams and Lotions

No. 15[11]

Dow Corning 200 Fluid, 100 cS	5.0
Mineral Oil, light	10.0
Arlacel 169 (*Atlas Chemical*)	3.0
Hystrene S-47 (*Humko*)	1.0
Beeswax, USP, bleached	0.5
Borax, cosmetic grade	0.5
Water	80.0
Methyl Parahydroxy-benzoate	*q.s.*
Perfume	*q.s.*
Color	*q.s.*

Mix all the ingredients except the perfume. Heat to between 165° and 170°F with high speed agitation (as with an Eppenbach mixer). Cool slowly to between 120° and 125°F while continuing agitation and add the perfume. Cool rapidly to room temperature and package.

No. 16[12]

Medicated Silicone Hand Cream

A.	Tegin 515	4.5
	Tegone A	1.5
	Stearyl Alcohol	2.0
	Stearic Acid	2.0
	Silicone Oil	2.0
	Lanfrax®	1.0
	Hexachlorophene	0.25
	Dilauryl Thiodipropionate	0.25
B.	Water	77.65
	Propylene Glycol	2.5
	Glycerin	2.5
	Allantoin	0.2
	Tegosept M	0.15
	Versene 100	0.05
C.	Water	3.0
	Sodium hydroxide	0.3
D.	Menthol	0.1
	Camphor	0.05

Heat A and B to 80°C, agitate enough to be certain that the Tegosept M is dissolved into the water phase and that the hexachlorophene is dissolved into the oil phase. Add C to B and then BC to A with agitation; cool to about 55°C with agitation; add D and cool to 45°C or below with agitation.

No. 17[12]

Tegamine P-13	0.3
Tegone A	6.0
Tegin 515	3.0
Dow Corning 200 Fluid	3.0
Spermaceti	2.0
Tegosept P	0.1
Water	80.25
Propylene Glycol	5.0
Citric Acid	0.2
Tegosept M	0.15

No. 18[13]
Silicone Hand Cream

A.	Stearic Acid	20.0
	Arlacel 60 Sorbitan Monostearate	1.5
	Tween 60 Polyoxyethylene Sorbitan Monostearate	3.5
	Silicone Fluid, 200 centistokes	5.0
	Sorbo Sorbitol Solution	20.0
B.	Water	50.0
	Sorbic acid (additional)	0.2
	Perfume	*q.s.*

Heat A to 70°C, B to 72°C. Add B to A slowly with thorough agitation and perfume at 50°C. Cool slowly, continuing agitation until cream sets up.

No. 19[13]

A.	Tegin	6.0
	Stearic Acid	8.0
	Petrolatum	6.0
	Tegosept P	0.1
B.	Water	63.25
	Glycerin	3.0
	Triethanolamine	1.5
	Tegosept M	0.15
C.	Talc	12.0

Heat B and A and add B to A with agitation. When uniform, add C, cool, and package.

Emollient Creams and Lotions

Emollient creams are designed to soften the skin and relieve its dry condition.

No. 20[14]

A.	Tegone A	15.0
	Tegamine P-13	0.3
	Silicone Oil	1.0
	Tegosept P	0.1
B.	Water	77.55
	Propylene Glycol	5.0
	Phosphoric Acid (10% solution)	0.9
	Tegosept M	0.15

Add A to B with agitation.

No. 21[14]

A.	Tegone A	6.0
	Iso-Lan	5.0
	Tegin 515	3.0
	Stearyl Alcohol	3.0
	Mineral oil (light N. F.)	15.0
	Peach kernel oil	5.0
	Isopropyl myristate	3.0
	Stearic acid	2.5
	Dilauryl thiodipropionate	0.05
	Tegosept P	0.1
B.	Water	50.95
	Propylene glycol	5.0
	Triethanolamine	1.2
	Versene 100	0.05
	Tegosept M	0.15

Heat A and B; add B to A with agitation; cool and package.

No. 22[15]
W/O Emollient Cream

A.	Petrolatum	35.0
	Mineral Oil	15.0
	Paraffin Wax	5.0
	Ceresin Wax	5.0
	Lanolin	1.0
	Arlacel 83	2.0
	Atlas G-1425 Lanolin Derivative	4.0
B.	Water	30.3
	Sorbo 70% Sorbitol Solution	2.5
	Magnesium Sulfate	0.2
	Preservative	*q.s.*
C.	Perfume	*q.s.*

Warm the oil phase A and water phase B separately to 70-75°C. Add B gradually to A while stirring. Add perfume at 55-60°C. Homogenize at 55-60°C, filling directly into jars.

No. 23[16]
Moisture Cream

A.	Isopropyl Myristate	2.0
	Mineral Oil 65/75	8.0
	Preservative	*q.s.*
	Stearic Acid (triple pressed)	4.0
	Cerasynt PA	7.0
	Spermaceti	2.0
	Ceraphyl 140-A	3.5
	Cerasynt 945	1.5
B.	Water, deionized	66.7
	Triethanolamine	0.3
	Propylene Glycol	5.0
C.	Perfume 0127	0.2

Heat A and B separately to 70-75°C. Add B to A with agitation. Cool to 40-45°C with stirring. Add C and continue stirring until thoroughly dispersed. Package.

No. 24[17]
Dry Skin Conditioning Lotion

Amerlate P	3.0
Amerchol L-101	4.0
Stearic Acid XXX	2.0
Glyceryl Monostearate, neut.	1.0
Mineral Oil, 70 vis.	15.0
Triethanolamine	0.8
Glycerine	5.0
Water	69.2
Perfume and Preservative	*q.s.*

No. 25[18]
Dry Skin Cream

Alcolan	25
Deodorized AAA Lanolin	5
Mineral Oil, 90 vis.	5
Water	65

Vanishing Creams

Vanishing creams are soft, highly emollient, nongreasy hand, face, and body creams, and are used as excellent makeup foundations and sunscreen vehicles.

No. 26[19]

		26a	26b
A.	Amerchol L-101	10.0	5.0
	Acetulan	5.0	----

Modulan	----	10.0
Solulan 98	3.0	3.0
Glyceryl Monostearate, neut.	5.0	5.0
Stearic acid XXX	15.0	15.0
Polyoxyl 40 Stearate, USP	3.0	3.0
B. Propylene Glycol	5.0	5.0
Water	54.0	54.0
C. Perfume and Preservative	*q.s.*	*q.s.*

Add B at 90°C to A at 90°C while mixing. Continue mixing and cool to 40°C. Add C and mix while cooling to 35°C.

No. 27[20]

Cutina® MD	4.0
Stearine	16.0
Eumulgin® C 700	3.0
Eutanol® G	3.0
White mineral oil	3.0
Triethanolamine	0.5
Water	70.5
Perfume and Preservative	to 100.0

No. 28[21]

A. Stearic Acid	25.0
Spermaceti	5.0
B. Amino-Methyl-Propanediol	1.5
Glycerine	8.0
Water	60.5

Heat B to about 75°C (167°F), while heating A in another container. When both mixtures are at the same temperature and homogeneous, slowly and thoroughly stir A into B. The temperature must be maintained at approximately 75°C throughout this operation. After A has been added, discontinue heating; however, maintain vigorous stirring until the mixture thickens. Slow the stirring and change to a kneading action. Like all stearate creams, it should be allowed to stand overnight and then be thoroughly remixed.

No. 29[22]

Stearic Acid	14.0
Triethanolamine	0.7
Du Pont *Lorol* 24 Cetyl Alcohol	3.0
Duponol C	0.5
Glycerine	5.0
Water	76.8

No. 30[23]

A. Crodafos N-3 Acid	4.0
Lanolin *Super Corona*	2.0
Propylparaben	0.1
Novol Cosmetic Oleyl Alcohol	1.5
Spermaceti	1.2
Stearic Acid (triple pressed)	8.5
B. Water	70.0
Methylparaben	0.1
Carbowax 4000 (Polyethylene Glycol-4000)	10.0
Triethanolamine	2.6

C. Perfume *q.s.*

Melt A and bring to 75°C. Heat B to the same temperature or slightly higher, and when both phases are dissolved, add B to A with mechanical agitation. Hot fill at 50°C after addition of C.

Cold Creams

No. 31[24]

Alcolan 36-W	5.0
Beeswax	10.0
Mineral Oil, 70 vis.	45.0
Paraffin (145°F)	5.0
Borax	0.7
Water	34.3

No. 32[25]

A.	Mineral Oil	50.0
	Beeswax	7.0
	Tween 40	2.0
	Atlas G-1726 Beeswax Derivative	8.0
B.	Water	33.0
	Preservative	*q.s.*
C.	Perfume	*q.s.*

Add A at 70°C to B at 70°C, slowly and with thorough agitation. Perfume at 50°C. Pour at 45-50°C.

No. 33, 34[25]

W/O Cold Creams

		33	34
A.	Petrolatum	31.0	35.0
	Mineral Oil	20.0	15.0
	Paraffin Wax (or Microcrystalline Wax)	7.0	5.0
	Lanolin	3.0	3.0
	Ceresin Wax	---	5.0
	Arlacel 83	4.0	4.0
B.	Water	32.3	30.3
	Sorbo 70% Sorbitol Solution	2.5	2.5
	Magnesium Sulfate	0.2	0.2
	Preservative	*q.s.*	*q.s.*
C.	Perfume		*q.s.*

Warm the oil phase A and water phase B separately to 70-75°C. Add the water phase gradually to the oil phase while stirring. Add perfume at 55° to 60°C. Homogenize at 55-60°C, filling directly into jars.

No. 35[26]

A.	Tegin P	6.0
	Protegin X	5.0
	Beeswax	16.0
	Ozokerite	5.0
	Mineral Oil	43.6
	Tegosept P	0.1
B.	Water	22.85
	Borax	1.3
	Tegosept M	0.15

Heat A and B until both are homogeneous, add B to A with agitation until smooth, cool and agitate, and package.

No. 36[27]

Amerchol L-101	4.00
Solulan 75	1.60
Tegin*	2.40
Stearic Acid	3.20
Mineral Oil, 80-90 vis.	9.60
Tegosept P*	0.08
Triethanolamine	0.40
Distilled Water	78.72

* *Goldschmidt Chem. Corp.*

To Aerosol

Above Concentrate	92.0
Genetrons 114/12† (43:57)	8.0

† *Gen. Chem. Div., Allied Chem. Corp.*

Cleansing Creams

No. 37[28]

A.	Beeswax	80
	Light Mineral Oil	490
	Paraffin	70
	Cetyl Alcohol	10
	Ethomeen C-25	10
B.	Water	334
	Borax	4
	Carbopol 934	2
	Color and Perfume	*q.s.*

Prepare B by completely dispersing the Carbopol 934 in the water and then blending in the borax. Heat both A and B to 75°C and add A to B with vigorous agitation. Once the mix has been blended to uniformity, cool it rapidly to 30°C. Perfume as desired when the blend reaches 50°C. **Note:** This formulation was prepared without homogenization and still gave a very white, smooth, effective product.

No. 38[29]

A.	Tegin P	6.5
	Mineral Oil	24.0
	Oleic Acid	3.0
	Lanolin	1.0
	Dilauryl Thiodipropionate	0.05
	Tegosept P	0.1
B.	Water	63.7
	Triethanolamine	1.5
	Tegosept M	0.15

Heat A and B and add B to A with agitation until AB is uniform. Cool and package.

No. 39, 40[30]

Enriched Cleansing Creams

Firm but light-textured glossy enriched creams for cleansing and moisturizing.

		39	40.0
A.	Amerchol L-101	3.0	3.0
	Amerlate P	1.0	1.0
	Beeswax, USP	10.0	17.0
	Mineral Oil, 70 vis.	43.15	21.5
	Isopropyl Palmitate	----	21.5
	Ozokerite	7.0	----
	Glyceryl Monostearate, s.e.	2.0	----
	Glyceryl Monostearate, neut.	----	2.0
B.	Triethanolamine	0.25	----
	Borax, USP	0.6	1.0
	Water	33.0	33.0
	Perfume and Preservative	*q.s.*	*q.s.*

Add B at 75°C to A at 75°C while stirring. Continue mixing and cool to 30°C.

No. 43[32]
W/O Cleansing Lotion

A.	Beeswax	10.0
	Mineral Oil	50.0
	Arlacel 83 Sorbitan Sesquioleate	1.0
	Lanolin	3.1
B.	Borax	0.7
	Water	35.2
	Sorbic Acid (additional)	0.2
C.	Perfume	*q.s.*

Heat A to 70-75°C; heat B to the same temperature. Add B to A gradually, while stirring. Perfume at 50°C and continue stirring until cool.

Night Creams

No. 44[33]
W/O Night Cream

Volpo 3	1.6
Super Hartolan (*Croda*)	1.6
Lanolin, Super Corona (*Croda*)	6.0
Cetyl Alcohol (*Croda*)	2.0
White Beeswax	10.0
Blandol, light mineral oil (*Sonneborn*)	5.0
Glyceryl Monostearate, pure	2.0
Sesame Oil	5.0
Deltyl Prime, Isopropyl Palmitate (*Givaudan*)	10.0
Propylparaben, USP	0.2
Butylated Hydroxy Anisole	0.2
Ceresine Wax, 160°F mp	3.6
Borax	1.0
Methylparaben, USP	0.1
Propylene Glycol	5.0
Distilled or Deionized Water	46.7
Perfume	*q.s.*

Add the water phase to the oils at 70°C, with mechanical agitation. Stir and cool to 40-50°C. Perfume and continue to cool to as low a temperature as possible before cream solidification. Homogenize at room temperature for better stability.

No. 45[34]

A.	Tegone A	6.0
	Tegin 515	3.0
	Iso-Lan	2.0
	Peach Kernel Oil	9.0
	Isopropyl Myristate	9.0
	Mineral Oil	9.0
	Petrolatum	9.0
	Stearic Acid	3.0
	Dilauryl Thiodipropionate	0.05
	Tegosept P	0.1
B.	Water	43.2
	Propylene glycol	5.0
	Triethanolamine	1.5
	Tegosept M	0.15

Heat A and B and add B to A with constant agitation, cool, and package.

No. 46[34]

A.	Tegamine P-13	0.3
	Tegone A	2.6
	Tegine 515	3.3
	Protegin X	3.5

	Iso-Lan	2.6
	Peach Kernel Oil	8.0
	Mineral Oil (light N. F.)	6.3
	Spermaceti	1.0
	Cetyl Alcohol	1.0
	Dow Corning 200 Fluid	1.0
	Dilauryl Thiodipropionate	0.05
	Tegosept P	0.1
B.	Water	64.2
	Propylene glycol	5.0
	Phosphoric acid (10% solution)	0.9
	Tegosept M	0.15

Heat A and B, add B to A with constant agitation, cool, and package.

No. 47[35]

Liquid Night Cream

Lanolin Absorption Base	24.0
Robane®	3.0
Isopropyl Myristate	5.0
Lanolin Oil	2.0
Carolate®	3.0
Propylene Glycol	5.0
Preservative	0.2
Water and Perfume	to 100.0

Makeup and Foundation Creams and Lotions

No. 48[36]

Liquid Makeup

A.	Stearic Acid (triple pressed)	3.5
	Mineral Oil No. 45*	9.0
	Ceraphyl 140-A	3.0
	Preservative	*q.s.*
	Cerasynt PA	3.0
B.	Water, deionized	60.9
	Triethanolamine	0.5
	Propylene Glycol	6.0
	Veegum HV†	0.4
C.	Sodium Lauryl Sulfate	0.5
	Kaloin No. 347††	2.3
	Alpine Talc No. 117††	4.3
	Titanium Dioxide No. 3328††	4.5
	Pigments††	2.0
	Calcium Stearate	2.0
D.	Perfume	0.2

Heat the water to 60°C and add the Veegum with agitation. Add the balance of B and continue agitation and heat to 70-75°C. Heat A to 70-75°C. When both A and B are at about the same temperature, add B to A with agitation. Continue agitation and cool to 50°C. Add C until completely uniform. At 40-45°C add D and stir until thoroughly dispersed. Mill and package.

* *Sonneborn Chemical and Refining Co.*
† *R. T. Vanderbilt Co., Inc.*
†† *Whittaker, Clark and Daniels, Inc.*

No. 49[36]

Cream Makeup

A.	Stearic Acid (triple pressed)	2.5
	Cerasynt PA	6.0
	Mineral Oil 65/75	8.0
	Ceraphyl 31	3.5
	Preservative	*q.s.*
B.	Water, deionized	61.2
	Triethanolamine	1.0
C.	Bentonite No. 50*	1.3
	Kaolin No. 347*	4.85

Talc No. 5211*	4.5
Titanium Dioxide No. 3328*	4.3
Pigments*	2.0
Sodium Lauryl Ether Sulfate	0.5
D. Perfume	0.25

Heat A and B separately to 70-75°C. Add A to B with agitation and continue stirring to 50-55°C. Add C and stir until the mixture is smooth and homogeneous. Cool to 45-50°C and add D. Continue stirring until thoroughly dispersed. Mill and package.

* *Whittaker, Clark and Daniels, Inc.*

No. 50[37]

Liquid Cream Makeup Base

Acetulan	4.0
Amerchol L-101	6.0
Solulan 98	2.0
Stearic Acid XXX	5.0
Glyceryl monostearate, neut.	2.5
Triethanolamine	1.0
Propylene Glycol	5.0
Water	74.5
Perfume and Preservative	*q.s.*

Add the water phase at 85°C to the oil phase at 85°C while stirring. Continue mixing and cool to room temperature. To prepare the finished makeup, add 10 parts solids to 90 parts of the base.

No. 51[38]

W/O Pigmented MakeUp Base

A. Gel Base		32.0
Arlacel 186	1 part	
Sorbo	9 parts	
B. Mineral Oil		10.0
Beeswax		1.5
Ceresin Wax		1.0
Preservative		*q.s.*
C. Titanium Dioxide		20.0
Water		35.5

Prepare the gel base A by adding Sorbo slowly to Arlacel while agitating the mixture. Blend B and A and heat to 70°C. Blend in C and mix until uniformly dispersed. Heat the water to 72°C and mix into ABC thoroughly, making sure all ingredients on the sides of the vessel are included. Stir until cool. Mill or homogenize for maximum smoothness and stability.

Face Cream

No. 52[39]

Stearyl Alcohol	3.5
Cetyl Alcohol	3.5
Sodium Lauryl Sulfate	1.0
Robane®	5.0
Stearic Acid	10.0
Glycerin	6.0
Triethanolamine	0.5
Borax	0.1
Preservative	0.2
Water and Perfume	to 100.0

No. 53[40]

O/W Face Cream

A. Stearic Acid	15.0
Isopropyl Myristate-	

Palmitate	1.0
Arlacel 60	2.0
Tween 60	1.5
B. Sorbo 70% Sorbitol Solution	3.0
Water	77.5
Preservative	*q.s.*
C. Perfume	*q.s.*

Add A at 90°C to B at 95°C. With rapid stirring, add C at 50°C. Stir occasionally to 25°C. Rework and pack next day.

No. 54[41]

This high gloss, pure white, rich textured, high-oil-content emollient washable cream can be used on all body areas. It is particularly suited for luxurious and enriched creams, baby products, hormone creams, makeup bases, eye creams, and cleansing cream products.

A. Amerchol L-101	5.0
Modulan	1.0
Spermaceti	4.0
Beeswax, USP	10.0
Mineral Oil, 70 vis.	20.5
Petrolatum, USP white	20.0
Arlacel 161*	3.0
Tween 60*	1.0
Polyethylene Glycol 400 Monostearate	2.0
Methocel 65HG, 4000 cP†	0.1
B. Propylene Glycol	5.0
Water	28.4
Perfume and Preservative	*q.s.*

Add B at 75°C to A at 75°C while stirring at moderate speed. Mix while cooling. At 40°C add the perfume. Continue mixing to 35°C and fill. The cream sets up gradually at room temperature. **Modifications:** Variations in consistency can be obtained by substituting Amerlate P or Waxolan for up to 3% of the mineral oil.

* *Atlas Chemical Industries*
† *Dow Chemical Co.*

No. 55[42]

An all purpose cream, excellent in appearance and functional properties, may be prepared as follows:

Ceralan	1.0
Beeswax, white USP	10.0
Paraffin 145°F	4.0
Mineral Oil, 70 vis.	50.0
Borax	0.7
Water	34.3

No. 56[43]

Dehydag® WaxO	2.0
Beeswax, white	3.0
Eumulgin® C 700	2.0
Stearic Acid	10.0
Eutanol® G	6.0
Lanolin, anhydrous	2.0
White Mineral Oil	10.0
White Petroleum Jelly	15.0
Triethanolamine	0.6
Water	49.4
Perfume and Preservative	*q.s.*

No. 57[44]

A. Tegin P	3.0
Tegin	2.0
Stearic Acid	5.0

	Oleic Acid	3.0
	Mineral Oil	3.0
	Cetyl Alcohol	1.75
	Lanolin	1.35
	Dilauryl Thiodipropionate	0.05
	Tegosept P	0.1
B.	Water	76.0
	Glycerin	3.0
	Triethanolamine	1.5
	Tegosept M	0.15
C.	Titanium dioxide	0.1

Heat A and B and add B to A with constant agitation. When thoroughly mixed, add C.

Anionic and Cationic Cream and Lotion Bases

No. 58[45]

Acid Cream

Du Pont *Lorol* 24 Cetyl Alcohol	15.0
Spermaceti	5.0
Duponol C	1.0
Lactic Acid (85%)	1.2
Water	72.8
Glycerine	5.0

Lorol 28 Stearyl Alcohol—USP grade—can be substituted for cetyl alcohol to produce a slightly heavier consistency. This cream is prepared by melting or dissolving the oil-soluble and water-soluble ingredients separately, heating both to about 176°F (80°C), and then adding the water phase to the oil phase with vigorous agitation. Dissolve the cetyl alcohol or stearyl alcohol in the oil phase and *Duponol* C in the water phase.

No. 59[46]

Cationic Lotion

An elegant lotion vehicle for ready incorporation of cation-active compounds is illustrated below. It is an ideal base for quaternary antiseptics, hair conditioners, and antistatic compounds, and is suggested for baby lotion, and general antiseptic hand, body, or hair lotion.

A.	Amerchol L-101	6.0
	Solulan 16	1.0
	Modulan	1.0
	Cetyl Alcohol	1.5
	Hyamine 10X*	0.1
	Emcol E-607S†	0.6
	Propylparaben	0.125
B.	Propylene Glycol	6.0
	Water	72.425
C.	Sodium Chloride	0.125
	Sodium Benzoate	0.125
	Water	11.0
Perfume		*q.s.*

Add B at 70°C to A at 70°C while mixing. Add C to AB and mix while cooling to 40°C. Add the perfume. Continue mixing and cool to room temperature.

* *Rohm and Hass*
† *Witco Chemical Co., Emulsol Div.*

No. 60[47]

Cationic Lotion

A.	Mineral Oil 65/75	9.0
	Solulan 75*	1.0
	Cerasynt W	3.5
	Isopropyl Linoleate	1.0

* *American Cholesterol Products, Inc.*

	Emcol E-607S†	0.5
	Ceraphyl 140-A	1.5
B.	Water, deionized	68.5
	Sorbitol	15.0

Heat A and B separately to 70-75°C. Add A to B with agitation, cool to 45°C, and package. **Note:** This lotion is intended for use with active ingredients requiring a cationic base.

† *Witco Chemical Co., Inc.*

Alcoholic Lotions

No. 61[48]

Polychol 5	30.0
Mineral Oil, 70 vis.	30.0
SD 40 Alcohol, anhydrous	20.0
Propylene Glycol	5.0
Water	15.0

No. 62[48]

Clear Alcoholic Hand/Body Lotion

A.	Crodafos N-3 Acid	2.5
	Novol Cosmetic Oleyl Alcohol	10.0
B.	Water	42.5
	Triethanolamine	1.0
95% Ethyl Alcohol		44.0

Dissolve the Crodafos in the Novol and bring to 45-50°C. Dissolve the Triethanolamine in the water and bring to 50°C. Add B to A and stir to produce a thick white emulsion. Stir and cool to about 30°C and then add the the alcohol. The solution will clear when almost all the alcohol has been added.

Special Purpose Creams and Lotions

No. 63[48]

Protein Conditioning Cream

Skliro	2.0
Polawax, regular	3.0
Novol	1.0
Petrolatum, White Protopet IS (*Sonneborn*)	4.0
Stearyl Alcohol (*Croda*)	4.0
Ammonyx 4 Stearyl Dimethyl Benzyl Ammonium Chloride (*Millmaster Onyx*)	10.0
Lanasan CL (*Sandoz Chemical Corp.*)	1.0
Glycerin	3.5
Water	71.3
Sorbic Acid	0.2
Perfume	*q.s.*

No. 64[48]

Special Enriched Skin Cream

Super Hartolan	1.8
Novol	2.0
Mineral Oil, light	20.0
Petrolatum, Alba Protopet	30.0
Super Corona Lanolin, USP	14.0
Beeswax, white	5.0
Titanium Dioxide, oil dispersible	0.3
Vitamin A Palmitate	0.1
Mixed Tocopherols	0.1
Apricot Kernel Oil (oil persic)	12.0
Butylated Hydroxy Anisole	0.1
Propylparaben, USP	0.2

Borax 0.15
Water 14.25
Perfume *q.s.*
FD and C Yellow No. 5
  Aqueous Solution *q.s.*

Disperse the TiO_2 in the mineral oil and add to the other oil phase ingredients with the exception of the vitamins. Bring the water phase and oil phase to 60°C, add the vitamins to the oils, then add the aqueous phase to the oils with mechanical agitation. Stir to 50°C and perfume. Add the color solution at any time. Pour at 44-46°C.

No. 65[49]
Nutritive Cream
(moderately oily)

Cutina® MD	18.0
Eumulgin® C700	4.0
Eutanol® G	20.0
Glycerine	5.0
Witch Hazel Extract, dist.	5.0
Boric Acid	0.2
Triethanolamine	1.0
Water	46.8
Preservative and Perfume	*q.s.*

No. 66[50]
W/O Zinc Oxide Cream

A.	Gel Base	30
	Arlacel 186	1 part
	Sorbo	9 parts
B.	Beeswax	1
	Ceresin Wax	1
	Mineral Oil	10
	Zinc Oxide, USP	20
	Water	38
	Preservative	*q.s.*

Prepare the intermediate gel base A by adding the Sorbo slowly to the Arlacel 186 with continuous agitation. Add B and heat to 70°C. Add the zinc oxide and mix until uniformly dispersed. Heat the water to 72°C and mix into the blend until it cools to room temperature using a vessel-fitting type blade. Milling is recommended to improve smoothness and shelf life.

No. 67[51]
Cream Sachet

A.	Tegacid Special	15.0
	Spermaceti	5.0
	Tegosept P	0.1
B.	Water	45.75
	Glycerin	5.0
	Tegosept M	0.15
C.	Perfume Oil	10.0
D.	Talc	15.0
	Titanium dioxide	4.0

Heat A and B and add B to A with constant agitation, cool, and add C and D in succession.

No. 68[52]
O/W Baby Lotion

A.	Mineral Oil	35.0
	Lanolin	1.0
	Cetyl Alcohol	1.0
	Arlacel 80	2.1
	Tween 80	4.9
	Hexachlorophene	0.5
B.	Water	55.5
	Preservative	*q.s.*
C.	Perfume	*q.s.*

Heat A to 60°C, B to 62°C.

Add B to A slowly with agitation. Add perfume at 50°C. Continue agitation to room temperature.

No. 69[53]

After Shave Collapsible Foam Aerosol

Polychol 40	1.5
Cetyl Alcohol (*Croda*)	1.5
SD 40 Alcohol (95%)	59.0
Hyamine 1622 (*Rohm and Haas Corp.*)	0.1
Menthol	0.1
Perfume	1.0
Water, distilled	36.8

To complete the aerosol, use 92% of the concentrate, 3% Propellant 11, 5% Propellant 12/114 Blend (40:60 ratio).

No. 70[54]

Day Cream

Cetyl Alcohol	3.5
Stearyl Alcohol	7.0
Sodium Lauryl Sulfate	2.0
Robane®	8.5
Sweet Almond Oil	5.0
Glycerin	5.0
Preservative	0.2
Water and Perfume	to 10b.0

Antiperspirant and Deodorant Creams and Lotions

No. 71[55]

Antiperspirant Roll-On

A.	Water, deionized	55.7
	Versene*	0.1
	Ceraphyl 140	2.0
	Cerasynt 945	5.0
	Veegum†	1.0
B.	Chlorohydrol 50%††	36.0
C.	Perfume 0125	0.2

* *The Dow Chemical Co.*
† *R. T. Vanderbilt Co., Inc.*

Heat the water to 60°C and add Veegum with agitation. Add the balance of A and continue agitation and heat to 70-75°C. Remove from heat and continue agitation to 45-50°C. Then add B very slowly so as not to "break" the emulsion. Continue stirring and cool to 40-45°C. Add C and continue agitation until thoroughly dispersed. Package.

†† *Reheis Co.*

No. 72[56]

Tegamine P-13	0.15
Tegin 515	1.5
Tegone A	1.5
Promulgen (type G)	0.3
Dow Corning 200 Fluid	0.25
Water	40.45
Cellosize QP-15000 (1% solution)	20.0
Propylene Glycol	5.0
Hydrochloric Acid (1 N)	0.7
Tegosept M	0.15
Chlorhydrol (50% solution)	30.0

No. 73[57]

Quick-Breaking Foam Antiperspirant

Polawax A-31	3.0
Anhydrous Alcohol	42.4
Water	26.0
Aluminum Chlorhydroxide	18.0
Hexachlorophene (*Sindar Corp.*)	0.1
Perfume	0.5

Propellent 12/114 (40:60 ratio)		10.0

No. 74[58]

O/W Deodorant Creams

A.	Stearic Acid	19.0
	Isopropyl Myristate-Palmitate	4.0
	Myrj 52	2.0
	Tween 60	8.0
	Hexachlorophene	0.5
B.	Water	66.5
	Preservative	*q.s.*
C.	Perfume	*q.s.*

Heat A to 70°C, B to 72°C. Add B to A slowly with agitation. Continue agitation until set up. Perfume and pack.

No. 75[59]

A.	Amerchol L-101	2.0
	Stearic Acid XXX	18.0
	Arlex*	2.0
	Mineral Oil, 70 vis.	2.0
B.	Propylene Glycol	7.0
	Amerdex†	3.0
	Water	64.5
	Triethanolamine	1.5
	Perfume, Muguet de Mai 233††	*q.s.*

Add B at 85°C to A at 85°C with stirring. Cool with stirring to 40°C and add the perfume. Continue mixing and cool to room temperature.

* *Atlas Chemical Industries*
† *Fleetwood Products*
†† *Roure Dupont Co.*

Bubble Baths and Bath Oils

No. 76[60]

Aerosol Bath Oil Foam

A.	Mineral Oil 345/355 (*Kaydol*)*	13.5
	Glycerine, USP	2.7
	Urea, crystalline	21.0
	Isopropyl Linoleate	1.6
	Foamole 2 APD	13.5
	Maprofix WAQ†	27.0
	Water, deionized	5.4
	Cerasynt IP	5.4
	Ceraphyl 140	4.3
	Lactic Acid 88%	4.0
B.	Perfume	1.6

Mix all the ingredients of A together and heat to 70-75°C. Remove from heat and cool with stirring to 40-45°C. Add B and continue stirring until thoroughly dispersed.

Aerosol:	Concentrate	75.0
	Propellent 12/114 (10-90)	25.0

* *Sonneborn Chemical and Refining Co.*
† *Onyx Chemical Co.*

No. 77[61]

Floating Bath Oil

Isopar Cosmetic Grade	40.0
Isopropyl Myristate	20.0
Enjay Hexadecyl Alcohol, cosmetic grade	30.0
Perfume Oil	9.0
Arlatone T	1.0

No. 78[62]

Bath Oil

Lanogel 61	2
Lanamine	10
Triethanolamine Lauryl	

Sulfate (45%)	16
Lauric Diethanolamide	10
Water	60
Perfume	2

No. 79[63]

Bubbling Bath Oil

	Parts by vol.
Orvus K Liquid	50
Isopropyl Palmitate	10
Water	40

Add the isopropyl palmitate to the Orvus K Liquid and mix until clear. Stir the water into this mixture until a clear product is obtained.

No. 80[64]

Bath Gel

A.	Foamole L	2.0
	Sarkosyl O*	1.0
	Water, deionized	20.0
	Sodium Lauryl Sulfate (WA Paste)	10.0
	Ammonyx LO†	1.5
	TEA Lauryl Sulfate (46% active)	53.5
	Foamole AR	9.0
B.	Perfume	3.0
	Color	*q.s.*

Heat A with slight agitation until clear—about 70-75°C. Remove from heat and continue agitation to 45-50°C. Add B and stir until thoroughly dispersed. Stop agitation and let entrapped air escape. Package.

* *Geigy Chemical Corp.*
† *Onyx Chemical Corp.*

Bubble Baths

No. 81[65]

Deriphat 170C	40.0
Alkanolamide (Ninol AA62)	1.2
Water	100.0

No. 82[66]

A.	Renex 690*	15.0
	Foamole 2 APD	3.0
	Sodium Lauryl Ether Sulfate	15.0
	Foamole L	7.0
	Water, deionized	60.0
B.	Perfume	0.5

Heat A with slight agitation until clear—about 70-75°C. Remove from heat and continue agitation to 45-50°C. Add B and stir until thoroughly dispersed. Package.

* *Atlas Chemical Industries, Inc.*

No. 83[67]

		83a	83b
A.	Volpo 3	5.0	5.0
	Volpo 10	5.0	–
	Fluilan, pure liquid Lanolin (*Croda*)	3.0	–
	Polychol 15, 15 mole PEG Lanolin Alcohol	3.0	–
	Triethanolamine Lauryl Sulfate (46% Al)	40.0	40.0
	Lauric Acid, 90% +	–	3.0
	Linoleic Diethanolamide, 1:1 adduct	–	2.0
	Crodafos N-3 Acid	–	2.0

Onyxol 345 (*Onyx Oil and Chemical Corp.*)	6.0	10.0
Propylene Glycol	–	10.0
Water	38.0	28.0
Perfume, Color, Additives	*q.s.*	*q.s.*

Warm A and stir until dissolved. When uniform, begin cooling and continue agitation; avoid excess aeration. Add the perfume at about 45°C.

Shampoos

No. 84[68]

Deriphat 160C	6
Triethanolamine (TEA) Salt* of Deriphat 170C	3
Triethanolamine Lauryl Sulfate (Trepenol WAT)	6
Ethoxylated Lanolin (Aureolan P, *American Lanolin Co.*)	4
Alkanolamide (Ninol AA62)	4
Water	77

* The TEA salt is prepared by adding an amount of triethanolamine equivalent to the acid value of the Deriphat:

$$\frac{\text{Acid value}}{56{,}100} \times 100 \times 149.2 = \text{grams TEA per 100 grams 170C}$$

No. 85[69]

Liquid Clear Shampoo

A. Sipon LT6 (American Alcolac)	35.0
Dow Corning 471 fluid	5.0
Perfume	0.1
Water	59.9

Mix A ingredients well; add water slowly to A with constant agitation.

No. 86[70]

Liquid Clear Low Eye-Sting Shampoo

Conco Sulfate WA	25.0
Condensate PE or PO	2.0
Miranol C2M	20.0
Hexylene Glycol	2.0
Polyoxyethylene Sorbitan Monolaurate	1.0
Water	50.0
Color, Perfume	*q.s.*

Blend all ingredients except perfume, color, and alkanolamide and heat to 70°C. Add alkanolamide and cool to 40-45°C. Add coloring and perfume.

No. 87[71]

Liquid Cream Shampoo Formulation for Normal Hair

A. Orvus WA Paste	50.0
P and G Amide No. 72	5.0
Zinc Stearate	2.0
B. Stearic Acid	0.4
Lanolin	1.0
Sodium Sulfate*	2.0
Water	39.6

Add A to warm water. Heat

* Added to thicker the solution. Viscosity can be increased as desired by increasing the Sodium Sulfate up to about 3%.

to approximately 160°F, then stir in B until it is completely melted. Allow to cool to room temperature while stirring. Disperse and mix the zinc stearate thoroughly with heavy stirring.

No. 88[72]
Special Shampoo for Oily Hair

Texapon® Q	30.0
Comperlan® KD	5.0
Dermolan	3.0
Aminosäuresol, type H (sol of amino acid)	1.0
Water	61.0
Preservative and Perfume	*q.s.*

No. 89[72]
Anti-Dandruff Shampoo

Texapon® Q	50.0
Comperlan® OD	5.0
Undecylene Acid Monoethanolamide	2.0
Hexachlorophene	1.0
Isopropyl Alcohol	2.0
Citric Acid	0.1
Water	39.9
Perfume	*q.s.*

No. 90[73]
Germicidal Shampoo and Hand Soap

Miranol CS Conc.	25-30
Lauric Acid Dialkanolamide 90%	3
Brij 30 (*Atlas*)	1
Hexylene or Propylene Glycol	1
Quaternary Germicide 50%	1
Water	69-64

No. 91[74]
Egg-Type Liquid Cream

Conco Sulfate WA	33.00
Condensate PO	4.00
Cerasynt IP	2.00
Powdered Egg	as required
1% Salt Solution	0.1-0.25
Alcolac HS-3	0.2-0.50
Water, Perfume, Coloring (D and C Yellow 10 seems to have good stability)	to 100.00

No. 92[75]
Clear Gel Shampoo

Volpo 20	5.0
Crodafos N-3 Neutral, DEA Oleyl Ether Phosphate	4.0
Onyxol 345, (*Millmaster Onyx Corp.*)	15.0
Lauric Acid	5.0
Triethanolamine Lauryl Sulfate (46% active)	20.0
Water	51.0
Perfume and Color	*q.s.*

Heat all ingredients except the perfume to 170°F and mix until clear. Cool to about 110°F and add perfume. Stir and cool until clear. Package before the gel stage (approximately 100°F) is reached.

No. 93[76]
Pearlescent Shampoo

A. Sodium Lauryl Sulfate (WA Paste)	35.0
Water, deionized	58.0
Solulan 98*	0.5
Cerasynt IP	3.0

* American Cholesterol Products, Inc.

Foamole L	1.0
Super Amide L-9†	2.5
B. Perfume 0120	0.2

Heat A with slight agitation until clear—about 70-75°C. Remove from heat and continue agitation to 45-50°C. Add B and stir until thoroughly dispersed. Package.

† Onyx Chemical Corp.

No. 94[77]

Professional Shampoo Concentrate

Miranol C2M conc.	50
Sodium Lauryl Sulfate (28% active)	20
Lauric Diethanolamide (high active)	20
Propylene Glycol	9
Sodium Xylene Sulfonate	1

Hair and After Shampoo Rinses

No. 95[78]

Cream Rinse Concentrate

A. Ammonyx 4: Stearyl Benzyl Dimethylammonium Chloride (*Onyx*)	10.0
Fluilanol	3.0
Stearyl Alcohol (*Croda*)	2.0
PEG 400 Distearate (*Kessler Div., Armour Inc.*)	3.0
Sodium Chloride	0.8
Water	81.2
Color and Perfume	*q.s.*

Melt A and heat to 65-70°C. Heat the water to 65-70°C and add to A. (Hold out some water to dissolve the salt.) Stir and cool to 40°C and add the salt solution, color, and perfume. Continue agitation and cool to about 32°C before filling.

No. 96[79]

Cream Rinse and Conditioner

A. Tegamine P-13	2.00
Tegamine S-13	1.00
Cetyl Alcohol	0.90
Propylene Glycol	9.00
Glycerin	9.00
Phosphoric Acid (85% solution)	0.95
B. Water	52.15
Natrosol 250H (1% solution)	25.00

Dissolve the phosphoric acid into some of the propylene glycol before A is heated. With agitation heat A and B to 70°C. With very good agitation, very slowly pour B into A. The clear solution that is formed will develop a creamy pearlescence upon standing.

No. 97[80]

This concentrated cationic hair rinse is easily diluted in water for use and is an effective conditioner after shampooing that leaves hair lustrous, soft, and manageable.

A. Amerchol L-101	5.0
Solulan C-24	2.0
Emcol E 607S*	2.0
Arlacel 165†	4.0
Cetyl Alcohol	1.0

* *Witco Chem. Co., Emulsol Div.*
† *Atlas Chemical Industries*

B.	Water	86.0
	Perfume and Preservative	*q.s.*

Add B at 70°C to A at 70°C while stirring. Cool with stirring to 40°C and add the perfume. Continue mixing and cool to 35°C. **Note:** Solulan 25 or Solulan 16 may be substituted for the Solulan C-24 to give slightly higher or lower viscosities.

No. 98[81]

A.	Triton X-400*	8.0
	Cerasynt 660	2.5
	Sodium Chloride	0.5
	Water, deionized	89.0
B.	Citric Acid	*q.s.* to pH 3.0-3.5
C.	Perfume	0.3
	Color	*q.s.*

Heat A to 70-75°C until homogeneous. Remove from heat, cool to 40-45°C, and add B. Continue agitation and add C until thoroughly dispersed. Package.

* *Rohm and Haas Co.*

Bleach Emulsions

No. 99[82]

Hydroquinone Bleach Cream

Tegone A	5.40
Tegacid Special	14.50
Iso-Lan	2.70
Mineral Oil	2.70
Silicone Oil	1.00
Tegosept P	0.10
Water	66.45
Propylene Glycol	5.00
Hydroquinone	2.00
Sodium m-Bisulfite	0.05
Sodium Lauryl Sulfate	0.10

No. 100[83]

Liquid Cream Bleach

Solulan C-24	3.00
Amerchol L-101	5.00
Cetyl Alcohol	2.50
Stearyl Alcohol	2.50
Phenacetin	0.04
Water	69.81
Hydrogen Peroxide, 35%	17.15
Perfume	*q.s.*

Cold-Wave and Neutralizer Emulsions

No. 101[84]

Cold-Wave Base

A.	Amerchol CAB	14.0
	Glyceryl Monostearate, neut.	6.9
	Span 20*	6.9
B.	Tween 20*	6.9
	Water	55.6
	Curling Agent (adjusted to pH 9.4 with ammonium hydroxide)	9.7

Add A at 60°C to B at 60°C. Cool with stirring to 40°C and homogenize. After the base AB has cooled, add the curling agent with pH adjusted to 9.4.

* *Atlas Chemical Industries*

No. 102[85]

Cold-Wave Neutralizer

Surfactol 365	1
Sodium Tripolyphosphate	1
Sodium Perborate, monohydrate	2
Sodium Bromate	4
Distilled Water	120

Perfume Oil	1

Hair Creams and Tonics

No. 103[86]

Cholesterol Hair Conditioning Cream

Polawax	9.0
Hartolan Lanolin Alcohols (30% cholesterol)	1.5
Propylaraben, USP	0.1
Lanolin Anhydr. USP. '*Super Corona*'	2.0
Novol Cosmetic Oleyl Alcohol	4.5
Cetyl Alcohol NF	5.0
Lecithin	0.2
Methylparaben	0.1
Water	77.6

No. 104[87]

Emulsified Pomade, O/W

A.	Amerchol L-101	2.50
	Modulan	2.50
	Solulan 16	1.00
	Petrolatum, USP white	25.00
	Paraffin Wax, m.p. 151	2.50
	Mineral Oil, 70 vis.	17.50
	Ucon 50HB 660*	5.00
B.	Carbopol 934†	0.25
	Water	41.75
	Sodium Hydroxide, 10% aq.	0.75
	Ethomeen C-25†† 10% aq.	1.25
	Perfume and Preservative	*q.s.*

Disperse the Carbopol in the water using high-speed agitation. Mix to form a uniform suspension free from lumps. Warm this suspension and A separately to 50°C. With mechanical agitation add to the water phase the sodium hydroxide solution followed by the Ethomeen solution. Add A to B and stir until emulsified. Cool with stirring to 40°C and add the perfume. Continue mixing and cool to room temperature.

Note: Consistency may be altered by adjusting the Carbopol content and/or the melting point of the oil phase.

* *Union Carbide Chemicals Co.*
† *B. F. Goodrich*
†† *Armour and Co.*

No. 105[88]

O/W Hair Dressing Cream

A.	Petrolatum	15.0
	Mineral Oil	10.0
	Lanolin	20.0
	Beeswax	12.0
	Tween 60 Polyoxyethylene Sorbitan Monostearate	5.0
	Arlacel 60 Sorbitan Monostearate	5.0
B.	Borax	1.0
	Water	32.0
	Sorbic Acid (additional)	0.2
C.	Perfume	*q.s.*

Heat A to 70°C, B to 72°C. Add B to A slowly with agitation. Perfume at 45°C and continue agitation until cool.

No. 106[89]

Hair Dressing

Sandrol 200 SN	2
Stearic Acid	1
Isopropyl Myristate	5
Lanolin, anhydrous	2
Light Mineral Oil	25

Water	65

No. 107[90]
Hair Cream, W/O

A.	Amerchol L-101	6.0
	Mineral Oil, 70 vis.	25.0
	Petrolatum, USP white	5.0
	Microcrystalline Wax, m.p. 170	5.0
	Aluminum Stearate No. 907*	1.0
	Sorbitan Sesquioleate	0.2
B.	Glycerine	5.0
	Water	52.8
Perfume and Preservative		*q.s.*

Add the aluminum stearate slowly to the mineral oil while mixing to prevent lump formation. Continue stirring and heat until clear and stringy. Add the remaining ingredients of A and maintain heat until wax is melted. Stir intermittently while cooling to 55°C. Add B at 55°C to A at 55°C with moderate stirring. Mix slowly while cooling to below 30°C. Remix one day later.

* *Whittaker, Clark and Daniels*

Note: The sorbitan sesquioleate concentration may be varied from 0.1 to 0.2% to alter the rate at which the cream liquifies on skin.

No. 108[90]
Cream Oil Hair Lotion, W/O

A.	Amerchol L-101	5.00
	Petrolatum, USP white	5.00
	Mineral Oil, 70 vis.	33.50
	Lanolin	3.00
	Sorbitan Sesquioleate	3.00
	Beeswax, USP	2.00
B.	Veegum*	0.25
	Borax, USP	0.50
	Water	47.75
Perfume and Preservative		*q.s.*

Disperse the Veegum thoroughly in the water at room temperature. Add B at 70°C to A at 70°C while stirring. Mix while cooling to 40°C. Add the perfume and continue mixing and cooling to room temperature. Homogenize.

* *R. T. Vanderbilt Co., Inc.*

No. 109, 110[91]
Hair Creams

		109	110
A.	Mineral Oil	50	128
	Lanolin	10	16
	Cetyl Alcohol	20	----
B.	Water	697	726
	Carbopol 934	5	5
	Propylene Glycol	212	117
	Triethanolamine	4	5
	Ethomeen C-25	3	3
	Propyl Parasept	1	1
	Perfume	*q.s.*	*q.s.*

Carefully disperse the Carbopol 934 in the blend of water, propylene glycol, and propyl parasept. Then stir in the triethanolamine and Ethomeen C-25. Heat both A and B to 75°C and add A to B with vigorous agitation. Stir the mix to uniformity, then cool rapidly to 30°C. Perfume should be added at 50°C.

No. 111[92]

Cream Hair Oil

Cream hair oils, which first appeared on the market during World War II, still retain their popularity and dominant position in the men's hair grooming preparation field.

Alcolan 40	14.25
Mineral Oil, 60 vis.	31.45
Water	54.00
Borax	0.30

SHAVING CREAMS

Aerosol Shaving Creams

No. 112[93]

A.	Stearic Acid, triple pressed	4.05
	Lauric Acid (*du Pont*)	1.35
	Dow Corning 470 Fluid	1.00
B.	Glycerine	9.00
	Triethanolamine (*Dow Chemical*)	3.24
	Water	71.36
Propellant 12		10.00
Perfume		*q.s.*

Heat A and B to 158°F. Add A to B and stir until cool. Add the perfum. Package in six-ounce lacquer-lined aerosol cans with foam nozzles. Add the propellant.

No. 113[94]

Modulan	0.8
Myristic Acid	2.0
Stearic Acid XXX	6.0
Cetyl Alcohol	0.5
Sorbo	3.0
Triethanolamine	4.0
Tween 20	2.5
Tween 80	2.5
Neutral Soap	1.0
Borax	0.1
Water	77.6
Perfume and Preservative	*q.s.*
Above concentrate	92.0
Genetrons 114A/12 (43:57)	8.0

Brushless Shaving Creams

No. 114[95]

A.	Stearic Acid	18.0
	Mineral Oil	5.0
	Tween 60 Polyoxyethylene Sorbitan Monostearate	5.0
B.	Sorbo Sorbitol Solution	5.0
	Borax	2.0
	Triethanolamine	1.0
	Water	64.0
	Sorbic Acid (additional)	0.2
C.	Perfume	*q.s.*

Heat A to 90°C, B to 95°C. Add B to A with agitation. After cream reaches the set point, stop mechanical agitation, add perfume, and stir occasionally by hand until the preparation reaches room temperature.

No. 115[96]

Tegin P	4.00
Stearic Acid	17.00
Petrolatum	2.00
Cocoa Butter	1.00
Lanolin	0.50
Tegosept P	0.10
Water	72.55
Glycerin	2.00
Triethanolamine	0.70
Tegosept M	0.15

Toothpastes

No. 116[97]

A. Sorbo (70% Sorbitol in water)	1000
Sodium Saccharine	6
Carbopol 934 Dispersion (6% in water)	600
B. Sodium Lauryl Sulfate	70
Water	484
Oil of Spearmint	40
C. Sodium Hydroxide (50% in water)	31
D. Dibasic Calcium Phosphate Dihydrate	1800

Dissolve B in one-third of the water. Mix A with the rest of the water. Blend A and B carefully to avoid foaming. Slowly add C to neutralize the Carbopol 934 while gently stirring. The solution will become thick as the neutralization progresses. Slowly mix in D. Stir until the particles are wetted out and all the lumps are broken up. Mill and deaerate the batch on conventional equipment. Add the flavoring during the last stages of milling to avoid evaporation loss. Tube the product.

No. 117[98]

A. Nipagin M	0.15
B. Texamid® 578L	1.50
C. Solution of Saccharine 1%	2.50
D. Texapon® K12	1.00
E. Toothpaste flavour	1.00
White Mineral Oil	2.00
Glycerine	20.00
F. Chalk	32.00
Water	39.85

Dissolve A in 36 parts of water by boiling. After cooling, soak B in this solution and add C. Dissolve D in the remaining amount of water (if necessary, by slight heating). Add F to E and then add ABC. While stirring carefully, add D and finally put the paste into a Vacuum Kneader until a smooth structure and bubble-free paste is obtained.

Suntan and Insect Repellant Creams and Lotions

No. 118[99]

W/O Sunscreen Preparation

A. Polychol 5	1.0
Volpo 3	3.0
Mineral Oil, 65-70 vis.	4.0
Robane, perhydrosqualene (*Robeco*)	8.0
Filtrosol A-1000 Sunscreen (*Schimmel Co.*)	2.0
B. Glycerin	4.0
Water, distilled	78.0

Dissolve A and cool to room temperature. Add B to A with vigorous mechanical agitation. No heat or homogenization is required.

No. 119[100]

Thin Suntan Lotion

A.	Mineral Oil	24.5
	Tween 60 Polyoxyethylene Sorbitan Monostearate	8.5
	Arlacel 60 Sorbitan Monostearate	1.5
	Benzyl Salicylate	2.0
	Benzyl Cinnamate	2.0
B.	Water	61.5
	Sorbic Acid (additional)	0.2

Heat A to 60°C, heat B to 62°C. Add B to A with thorough, but gentle stirring. After B has been added to A, permit the mixture to cool to room temperature with gentle stirring.

No. 120[100]

O/W Suntan Lotion

A.	Benzyl Salicylate	2.0
	Benzyl Cinnamate	2.0
	Mineral Oil	24.5
	Arlacel® 60 Sorbitan Monostearate	1.5
	Tween 60 Polyoxyethylene Sorbitan Monostearate	8.5
	Sorbic acid (preservative)	0.2
B.	Water	61.5

Heat A to 60°C, B to 62°C. Add B at 62°C to A at 60°C with vigorous agitation, being careful not to incorporate air into the mixture. Stir and allow the emulsion to cool to room temperature.

No. 121[100]

O/W Suntan Cream

A.	Monoglyceryl Para-Aminobenzoate	2.5
	Stearyl Alcohol	5.0
	Mineral Oil	26.0
	Arlacel 161 Glycerol Monostearate (non-self-emulsifying)	5.2
	Myrj 59 Polyoxyethylene Stearate	4.8
	Sorbic Acid (preservative)	0.2
B.	Water	56.5

Heat B to 62°C. Add B at 62°C to A at 60°C with vigorous agitation, being careful not to incorporate air. Allow the emulsion to cool while stirring to room temperature.

No. 122[101]

A.	Cerasynt WM	2.00
	Propylene Glycol	6.00
	Cetyl Alcohol N.F.	1.00
	Escalol 106	2.00
	Emulsynt 1060	5.00
	Stearic Acid (triple pressed)	2.00
	Peanut Oil	4.50
	Preservative	*q.s.*
B.	Triethanolamine	0.50
	Urea, crystalline	2.00
	Water, deionized	74.75
C.	Perfume	0.25

Heat A and B separately to 70-75°C. While stirring A, slowly add B and stir to 40-45°C. Add C and continue stirring until thoroughly dispersed. Package.

References

1. Commercial Solvents Corp.
2. B. F. Goodrich Chemical Co.
3. Atlas Chemical Industries, Inc.
4. Robeco Chemicals, Inc.
5. American Cholesterol Products, Inc.
6. B. F. Goodrich Chemical Co.
7. Atlas Chemical Industries, Inc.
8. Robeco Chemicals, Inc.
9. Croda, Inc.
10. American Cholesterol Products, Inc.
11. Dow Corning Corp.
12. Goldschmidt Chemical Corp.
13. Atlas Chemical Industries, Inc.
14. Goldschmidt Chemical Corp.
15. Atlas Chemical Industries, Inc.
16. Van Dyk and Co.
17. American Cholesterol Products, Inc.
18. Robinson Wagner Co., Inc.
19. American Cholesterol Products, Inc.
20. Henkel International GMBH, A. H. Carnes Co., agent
21. Commercial Solvents Corp.
22. E. I. du Pont de Nemours and Co.
23. Croda, Inc.
24. Robinson Wagner Co., Inc.
25. Atlas Chemical Industries, Inc.
26. Goldschmidt Chemical Corp.
27. American Cholesterol Products, Inc.
28. B. F. Goodrich Chemical Co.
29. Goldschmidt Chemical Corp.
30. American Cholesterol Products, Inc.
31. Van Dyk and Co.
32. Atlas Chemical Industries, Inc.
33. Croda, Inc.
34. Goldschmidt Chemical Corp.
35. Robeco Chemicals, Inc.
36. Van Dyk and Co.
37. American Cholesterol Products, Inc.
38. Atlas Chemical Industries, Inc.
39. Robeco Chemicals, Inc.
40. Atlas Chemical Industries, Inc.
41. American Cholesterol Products, Inc.
42. Robinson Wagner Co., Inc.
43. Henkel International GMBH, A. H. CarnesCo., agent
44. Goldschmidt Chemical Corp.
45. E. I. du Pont de Nemours and Co.
46. American Cholesterol Products, Inc.
47. Van Dyk and Co.
48. Croda, Inc.
49. Henkel International GMBH, A. H. Carnes Co., agent
50. Atlas Chemical Industries, Inc.
51. Goldschmidt Chemical

Corp.
52. Atlas Chemical Industries, Inc.
53. Croda, Inc.
54. Robinson Wagner Co., Inc.
55. Van Dyk and Co.
56. Goldschmidt Chemical Corp.
57. Croda, Inc.
58. American Cholesterol Products, Inc.
59. Atlas Chemical Industries, Inc.
60. Van Dyk and Co.
61. Atlas Chemical Industries, Inc.
62. Robinson Wagner Co., Inc.
63. Proctor and Gamble, Co.
64. Van Dyk and Co.
65. General Mills
66. Van Dyk and Co.
67. Croda, Inc.
68. General Mills
69. Wyandotte Chemicals Corp.
70. Continental Chemical Co.
71. Proctor and Gamble, Co.
72. Henkel International GMBH, A. H. Carnes Co., agent
73. Miranol Chemical Co., Inc.
74. Continental Chemical Co.
75. Croda, Inc.
76. Van Dyk and Co.
77. Miranol Chemical Co., Inc.
78. Croda, Inc.
79. Goldschmidt Chemical Corp.
80. American Cholesterol Products, Inc.
81. Van Dyk and Co.
82. Goldschmidt Chemical Corp.
83. American Cholesterol Products, Inc.
84. R. E. Saute, G. J. Sperandio, and H. G. De Kay, *J. Soc. Cosmetic Chemists*, 4:270 (1953).
85. Baker Castor Oil Co.
86. Croda, Inc.
87. American Cholesterol Products, Inc.
88. Atlas Chemical Industries, Inc.
89. Clintwood Chemical Co.
90. American Cholesterol Products, Inc.
91. B. F. Goodrich Chemical Co.
92. Robinson Wagner Co., Inc.
93. Chevron Chemical Co., Oronite Division
94. American Cholesterol Products, Inc.
95. Atlas Chemical Industries, Inc.
96. Goldschmidt Chemical Corp.
97. B. F. Goodrich Chemical Co.
98. Henkel International GMBH, A. H. Carnes Co., agent
99. Croda, Inc.
100. Atlas Chemical Industries, Inc.
101. Van Dyk and Co.

chapter 5

EMULSION POLYMERIZATION

Emulsion polymerization is both one of the oldest and one of the newest forms of polymerization. Natural rubber is produced inside the rubber tree by a type of emulsion polymerization. Many synthetic rubbers and resins are emulsion polymerized today. The emulsion polymerization of floor polish polymers, for instance, accounts for at least a $12 million business annually.[5] Bovey *et al*[6] listed eight polymers that are made on a large scale by emulsion polymerization techniques: polystyrene, butadiene–styrene, butadiene–acrylonitrile, neoprene, polyvinyl chloride, vinyl chloride–vinyl acetate, vinyl chloride–vinylidene chloride, and fluorocarbons.

The essential advantages of emulsion polymerization include rapid polymerization of a monomer to a high molecular weight, use of a low-cost diluent, the low viscosity of the reaction mass, and excellent uniform heat transfer to and from the polymer. However, the major disadvantage of emulsion polymerization is that the bulk phase of the polymer is contaminated with the surfactant.[7]

Emulsion polymerization proceeds by a free-radical mechanism. The polymerization of the monomer does not occur within the emulsified particles as one might expect. ". . . [the] principle locus for the formation of the polymer particle is the monomer solubilized in the micelles. After polymerization has taken place, the polymer particle is ejected from the micelle and exists in the aqueous phase as a discrete particle surrounded and stabilized by an adsorbed layer of the . . . emulsifying agent."[8] Other authors disagree as to the exact location of the free-radical initiation of polymerization. The course of emulsion polymerization is often followed by the measured changes in the surface tension or other effects that result from a decrease in the concentration of micelles in the emulsion.

A typical emulsion polymer is composed of the following:

Internal phase	
Monomer	e.g. vinyl chloride
Initiator	e.g. potassium persulfate
Chain modifier or transfer agent to limit or control the molecular weight	e.g. alkyl mercaptan
Surfactant	e.g. sodium soap
Continuous phase	e.g. water

Acrylic Ester Homo- and Copolymers

Formula No. 1[1]

Acrylic Ester Homopolymer

1. Water	130.00
2. Abex 18S	14.00
3. Methyl Acrylate	100.00
4. Potassium Persulfate	0.20
5. Sodium Bisulfite (2% solution)	10.00

Procedure

a. Charge the water (1) into a reaction flask and purge with nitrogen for 15 minutes.

b. Add the Abex 18S (2) and dissolve.

c. Add 20% of methyl acrylate (3) to the flask and purge with nitrogen for another 5 minutes.

d. Heat the resulting emulsion to 45-50°C.

e. Add the potassium persulfate (4), allow to dissolve for 1-2 minutes, and then add 10% of the bisulfite solution (5).

f. After the reaction has started and a temperature rise of 5°C is observed, start addition of the remaining methyl acrylate and sodium bisulfite solution concurrently.

g. Maintain a reaction temperature of 70-75°C throughout the reaction cycle.

h. After the monomer addition is completed, raise the batch temperature to 80°C and maintain for 1 hour.

i. Cool to below 30°C and store.

No. 2[1]

1. Water	85.00
2. Abex 18S	14.00
3. Ethyl Acrylate	70.00
4. Methyl Methacrylate	29.00

5.	Methacrylic Acid	1.00
6.	Potassium Persulfate	0.40
7.	Sodium m-Bisulfite (2% solution)	15.00
8.	Sodium Hydroxide (5% solution)	10.00

Procedure

a. Charge the water (1) into a reaction flask and purge with nitrogen for 15 minutes.

b. Add the Abex 18S (2) and dissolve.

c. Add 20% of premixed monomers (3,4,5). Purge with nitrogen for another 5 minutes.

d. Heat the contents of the flask to 50°C and add the potassium persulfate (6).

e. To initiate the reaction, add 10% of the sodium bisulfite solution (7).

f. As soon as the reaction is started and after an exothermic temperature rise of 5°C is noticed, start addition of the remainder of the monomers and bisulfite solution concurrently.

g. Maintain the reaction temperature at 72-76°C throughout the addition period of the monomers. After addition is completed, heat the batch to 85°C and maintain this temperature for ½ hour.

h. Cool the batch to room temperature and adjust the pH to 8-9 with the sodium hydroxide solution (8).

No. 3[1]

Vinyl Acetate–Acrylic Ester Copolymer

1.	Water	230.00
2.	Abex 18S	42.00
3.	Vinyl Acetate	175.00
4.	2-Ethylhexyl Acrylate	75.00
5.	Ammonium Persulfate	1.00
6.	Sodium m-Bisulfite (2% solution)	50.00

Procedure

a. Charge the water (1) into a reaction flask and purge with nitrogen gas for 15 minutes.

b. Charge the surfactant (2) and 20% of the monomer mixture (3,4).

c. Raise the temperature to 55°C and purge with nitrogen gas for another 5 minutes.

d. Add the catalyst (5) and 10% of the activator (6).

e. As soon as the reaction is exothermic, start the flow of the remaining monomer mixture and activator.

f. Maintain the reaction temperature at 60-65°C throughout the reaction.

g. After completion, heat to 70-75°C and hold for 1 hour.

h. Cool and store.

No. 4[1]

High Molecular Weight Acrylic Ester–Styrene Copolymer

Pre-Emulsion

1.	Water	109.00
2.	Abex 18S	10.00
3.	Tergitol Nonionic NPX	6.00
4.	Styrene	150.00
5.	2-Ethyl Hexyl Acrylate	150.00
6.	Methacrylic Acid	3.00
7.	Water	90.00
8.	Abex 18S	16.00
9.	Potassium Persulfate	1.20
10.	Sodium m-Bisulfite (2% solution)	20.00

Procedure

a. Charge the water (7) and flush with nitrogen gas for 15 minutes.

b. Add the Abex 18S(8) and potassium persulfate (9).

c. Heat to 75-77°C and start the flow of pre-emulsion.

d. Preparation of pre-emulsion: Charge the water (1), Abex 18S(2), and Tergitol Nonionic NPX (3). Flow in the monomers (4, 5, 6) and emulsify thoroughly.

e. Add pre-emulsified monomers over a 1 to 1½ hour period.

f. When all the monomers have been added, start the flow of sodium metabisulfite solution (10) over a 30 minute period, maintaining the temperature between 75 and 78°C.

g. Upon completion, heat to 78-82°C and hold for another 45-60 minutes.

h. Cool to below 30°C and store.

No. 5[1]

Low Molecular Weight Acrylic Ester-Styrene Copolymer

1.	Water	100.00
2.	Abex 18S	14.00
3.	Ethyl Acrylate	69.00
4.	Styrene	30.00
5.	Methacrylic Acid	1.00

6. Potassium Persulfate	0.40
7. Sodium m-Bisulfite (2% solution)	15.00

Procedure

a. Add water to the reaction flask and purge with nitrogen gas for a period of 15 minutes.

b. Dissolve the emulsifying agent in the water and add 20% of the monomer mixture; stir at 100-200 rpm to obtain a uniform emulsion. Purge the emulsion with nitrogen gas for another 5 minutes.

c. Raise the temperature of the batch to 45-50°C. Add the potassium persulfate and allow to dissolve while stirring. Initiate the reaction by adding 1 ml of the 2% bisulfite solution.

d. Within a few minutes the temperature of the batch will rise. At 60°C, start a uniform flow of the remaining monomer mixture. Delay the addition over a 1 hour period. Add the 2% bisulfite solution concurrently at such a rate that by the time the monomer addition is completed only $\frac{1}{2}$ of all the bisulfite solution is consumed. Maintain the reaction temperature during the addition time at 70-75°C. Since, at the described monomer addition rate, the reaction is strongly exothermic, continuous moderate cooling is necessary to maintain the specified reaction temperature.

e. After monomer addition is completed, the reaction will continue to be exothermic for another few minutes. When the exotherm has ceased, heat the batch to 80°C and maintain this temperature for 30 minutes to 1 hour. During this heating cycle, the remaining half of the bisulfite solution is added at a uniform rate.

f. The batch is cooled to room temperature after the heating cycle is completed and neutralized to a pH of 8-9 with a preferred base.

No. 6[2]

Homopolymer Vinyl Acetate Emulsion for Adhesives

The following recipe can be adapted to the batch method of preparation. The resultant emulsion, stabilized with partially-hydrolyzed polyvinyl alcohol, has excellent solvent tolerance and mechanical stability.

Initial Charge	
Emulsion Heel (Vinac XX-210)	400.0
Distilled Water	37.0
Formopon	0.5

Delayed Charge	
Vinol PA-5 (13.8% solution)	145.0
Vinol 523 (10% solution)	55.0
Hydrogen Peroxide, 3%	25.0
Distilled Water	87.0
Sodium bicarbonate	0.9
Monomer Charge	
Airco H-Grade Vinyl Acetate	425.0
Surfynol PC	2.0
Reducing-Agent Solution	
Formopon	1.5
Distilled Water	35.0

Procedure

a. Prepare the initial charge by dissolving the Formopon in water and adding the solution to the emulsion heel (Vinac XX-210 is the recommended emulsion heel). Heat the charge to 70°C by means of a water bath.

b. Add the delayed charge, monomer charge, and reducing-agent solution—evenly and simultaneously—over a period of 3-3½ hours. During this time, allow the reaction temperature to climb to 80°C and hold at that level for the remainder of the polymerization.

c. When the delayed charge, monomer charge, and reducing agent solution have been added, maintain the temperature of the water bath at 80°C for 1 hour to reduce the free monomer content to 0.5% or less.

d. Cool the emulsion to room temperature with stirring, and transfer to a storage container.

Airco H-Grade—vinyl acetate monomer; Surfynol PC:—organic surface-active blend; Vinac XX-210:—homopolymer vinyl acetate emulsion; Vinol 523: medium-viscosity, partially-hydrolyzed polyvinyl alcohol; Vinol PA-5:—low-viscosity, partially-hydrolyzed polyvinyl alcohol *(Air Reduction Chemical and Carbide Company)*. Formopon:—reducing agent *(Rohm and Haas Co.)*.

Polyvinyl Acetate Homopolymers

No. 7[3]

1. Water	84.00
2. Abex 18S	17.00
3. Vinyl Acetate	100.00
4. Potassium Persulfate	0.40
5. Sodium Bisulfite (2% solution)	20.00

6. Sodium Hydroxide (5% solution)	as required to adjust pH to desired level

Procedure

a. Add the water (1) to a reaction flask and purge with nitrogen for 15 minutes.
b. Add the Abex 18S (2) and dissolve.
c. Add 25% of the monomer (3) and emulsify.
d. Purge with N_2 gas for another 5 minutes, then raise the temperature to 50-55°C.
e. Add the catalyst (4), allow 1-2 minutes for it to dissolve, then add 10% of the activator (5).
f. As soon as a 5°C increase is noticed, start delayed addition of the remaining monomer and bisulfite solution concurrently over 1-1½ hours.
g. Maintain the reaction temperature at 62°-65°C throughout the reaction cycle.
h. After monomer addition is completed, raise the temperature to 72°C and maintain for 1 hour.
i. Cool the batch to room temperature and adjust the pH with sodium hydroxide (6).

No. 8[3]

Initial Charge	
1. Water	130.00
2. Abex 22S	34.00
3. Vinyl Acetate	30.00
Pre-Emulsion	
4. Water	150.00
5. Itaconic Acid	3.00
6. Abex 22S	10.00
7. Vinyl Acetate	270.00
8. Potassium Persulfate	0.60
9. Sodium Bisulfite (3% solution)	30.00
10. Sodium Hydroxide (5% solution)	as required to adjust pH of latex to desired level

Procedure

a. Charge the water (1) into a reaction flask and purge with nitrogen gas for 15 minutes at the rate of 200 bubbles/min.

b. Add the Abex 22S (2) and disperse. Add the vinyl acetate (3). Purge with nitrogen gas for another 5 minutes before initiating reaction.

c. Prepare a pre-emulsion with items 4,5,6, and 7. First dissolve the itaconic acid in the water; then add the surfactant and vinyl acetate. Agitate at high speed for 5 minutes. (The resulting emulsion is stable for several hours and does not separate.) Add the emulsion to a graduated addition funnel connected to the reaction flask.

d. Raise the temperature of the contents of the reaction flask to 55°C and add the potassium persulfate (8) as is, and 5 cc of bisulfite solution (9). The reaction should start within a few minutes.

e. Allow the reaction temperature to rise to 66°C, then start the flow of pre-emulsion. (At this point the batch should not be refluxing. If this occurs because of rapid reaction rates and slow heat dissipation, the addition should be postponed until the reflux has subsided. Conversion of most of the initial charge into a polymer seed is essential for preventing inhibition upon the addition of the itaconic-acid-containing emulsion.)

f. Start the flow of the remaining bisulfite solution concurrently with the pre-emulsion.

g. Maintain the reaction temperature at 66°-70°C throughout the reaction. Initially the flow rate of the pre-emulsion should be slow (40-60 drops/min). As the reaction proceeds, the rate can be increased to 300-400 drops/min. Total addition time is approximately 1 hour.

h. After the reaction is completed, raise the temperature to 75°-80°C for 1 hour, then cool and adjust the pH to the desired level with 5% sodium hydroxide solution (10).

No. 9[3]

Vinyl Acetate–Dibutyl Maleate (Furmarate) Copolymer

1.	Abex VA40	16.00
2.	Water	225.00
3.	Potassium Persulfate	1.00
4.	Dibutyl Maleate (or Fumarate)	80.00
5.	Vinyl Acetate	320.00
6.	1% Sodium Sulfoxylate Formaldehyde Solution	25.00
7.	2% Potassium Persulfate	50.00
8.	Cellosize WP4400	1.75
9.	Sodium Bicarbonate	1.00

Procedure

a. Add the water and Cellosize WP4400 to a reaction flask and dissolve the Cellosize WP4400.
b. Purge with nitrogen gas for 15 min at a rate of about 200 bubbles/min.
c. Add the Abex VA40 (1), sodium bicarbonate (9), and potassium persulfate (3).
d. After dissolving the ingredients, purge with nitrogen gas for another 5 minutes.
e. Add 20% of the pre-mixed monomers, and with a slow purge (50 bubbles/min) of nitrogen gas heat to 55-57°C.
f. Add the reducing agent (6) over a 10-min period.
g. The reaction will become exothermic in a maximum of 2-3 minutes. At 66-67°C, start incremental addition of the monomers and catalyst solution concurrently over a 4 to 4½ hr period.
h. Keep the reaction temperature between 78-82°C.
i. At the end of the addition, let exotherm dissipate.
j. Heat to 85-87°C and hold for 1 hr.
k. Cool to below 30°C and store.

No. 10[3]
Alkali-Soluble Polymers

1.	Water	150.00
2.	Abex 22S	32.00
3.	Ethyl Acrylate	30.00
4.	Methacrylic Acid	10.00
5.	Water	100.00
6.	Abex 22S	12.00
7.	Ethyl Acrylate	150.00
8.	Methacrylic Acid	50.00
9.	Ammonium Persulfate	2.40
10.	Sodium m-Bisulfite (4.8% solution)	50.00

Procedure

a. Add items 1,2,3, and 4 to a reaction flask in that order, allowing for the Abex 22S to dissolve before proceeding with the addition of the monomers. Provide adequate stirring to form an emulsion.
b. Prepare an emulsion with items 5,6,7, and 8, using a high-speed stirrer. Place this emulsion in a graduate funnel that can be connected to the reaction flask.

c. Add the ammonium persulfate to the reaction flask and allow 1 min for it to dissolve. Then add 10 cc of bisulfite solution, and when exothermic start the flow of pre-emulsion and 40 cc of sodium bisulfite solution. Delay the addition over a period of 1½—2 hr.
d. Maintain the reaction temperature between 65-67°C.
e. After completing the addition of the pre-emulsion, heat the batch to 85°C and hold for ½ hr, then cool to room temperature.

No. 11[3]

Styrene–Butadiene Copolymer

1.	Abex AA65	9.25
2.	Water	126.00
3.	Butadiene	40.00
4.	Styrene	60.00
5.	$K_2S_2O_8$	1.00
6.	Sodium Toluene Sulfonate	0.30

Procedure

The experimental work was done on a laboratory scale using 32-oz soda bottles and utilizing a "pop-bottle reactor." Water and the nongaseous monomers were added to the bottles. The bottles were shaken well and after a short period in cold storage, potassium persulfate and butadiene monomers were added. The bottles were capped and placed into the pop-bottle reactor and allowed to thaw at 10-20°C for ½ hr.

Upon completion of the thawing operation, the water bath was increased to reaction temperature. The temperature and reaction cycles utilized were either 16 hr at 50°C or 6 hr at 50°C, followed by 2 hr at 70°C. Either method resulted in an adequate conversion of monomers.

No. 12[4]

Standard GR-S Rubber

Butadiene	75.0
Styrene	25.0
Water	180.0
Soap	5.0
n-Dodecyl Mercaptan	0.5
Potassium Persulfate	0.3

Polymerize at 50°C.

References

1. Alcolac Chemical Corp.
2. Air Reduction Chemical and Carbide Co.
3. Alcolac Chemical Corp.
4. F. A. Bovey, I. M. Kolthoff, A. I. Medalia, and E. J. Meehan, *Emulsion Polymerization*, p. 22, Interscience, New York, 1955.
5. Richardson Courts Floor Polish Market, *Chem. Engr. News*, May 16, 1966, p. 28.
6. Bovey et al, *op. cit.* p. 22.
7. *Ibid.* p. 17.
8. J. L. Molliet, B. Collie, and W. Black, *Surface Activity*, 2nd ed. p. 344, D. Van Nostrand Co., Inc., Princeton, N. J., 1961.

chapter 6

FOAMS AND ANTIFOAMS

RIGID POLYURETHANE FOAMS

Aromatic Base Polyols

Formula No. 1[1]

LK 380 (*Union Carbide*)	40.00
Quadrol (*Wyandotte*)	3.75
Refrigerant 11B	14.30
DBTDL (dibutyltindilaurate)	0.30
Dow Corning® 193 Surfactant	0.50
Nacconate 4040 (*Allied*)	35.00

No. 2[1]

Actol 52-460 (*Allied*)	100.00
Refrigerant 11B	38.00
Quadrol (*Wyandotte*)	8.00
Dabco 33LV (*Houdry*)	0.90
DMEA (dimethylethanolamine)	0.50
DBTDL (dibutyltindilaurate)	0.02
Dow Corning 193 Surfactant	1.50
Nacconate 4040 (*Allied*)	107.00

Chlorinated Polyester Base Polyols

No. 3[1]

HLR 250 (*Hooker*)	100.0
Refrigerant 11B	40.0
TEA (triethylamine)	2.0
TMG (tetramethylguanidine)	2.0
Dow Corning 193 Surfactant	2.0
PAPI (*Upjohn*)	99.6

No. 4[1]

Hetrofoam 250-280 (*Hooker*)	50.0
Dow Corning 193 Surfactant	0.5
Nacconate 4040 (*Allied*)	27.5

Methyl Glucoside Base Polyols

No. 5[1]

G-435 DM (*Olin*)	60.0
Vircol 82 (*Virginia-Carolina*)	40.0
Refrigerant 11B	40.0
TMBDA (tetramethylbutanediamine)	1.0
TMG (tetramethylguanidine)	0.5
Dow Corning® 193 Surfactant	1.5
PAPI (*Upjohn*)	88.0

No. 6[1]

G-435 DM (*Olin*)	60.0
204S (*Wyandotte*)	40.0
Refrigerant 11B	40.0
TMBDA (tetramethylbutanediamine)	1.0
TMG (tetramethylguanidine)	0.5

Dow Corning 193 Surfactant	1.5
PAPI (*Upjohn*)	110.0

No. 7[1]

Pentaerythritol Base Polyol

PEP 450 (*Wyandotte*)	64.50
Refrigerant 11B	33.50
TMBDA (tetramethyl-butanediamine)	0.50
DBTDL (dibutyltindilau-rate)	0.15
Dow Corning® 193 Surfactant	1.00
PP-130 (*Phelan*)	105.00

Sorbitol Base Polyols

No. 8[1]

G-2571 (*Atlas*)	43.6
Refrigerant 11B	16.0
DMEA (dimethylethanol-amine)	0.5
DBTDL (dibutyltindilaurate)	0.5
Dow Corning 193 Surfactant	0.5
Nacconate 4040 (*Allied*)	38.9

No. 9[1]

G-2410 (*Atlas*)	42.7
Refrigerant 11B	15.0
DMEA (dimethylethanol-amine)	0.5
DBTDL (dibutyltindilaurate)	0.6
Dow Corning 193 Surfactant	0.5
Nacconate 4040 (*Allied*)	32.6

Sucrose Base Polyols

No. 10[1]

SA 1049 (*Dow*)	142.10
Refrigerant 11B	56.20
DMEA (dimethylethanol-amine)	1.53
Dabco 33LV (*Houdry*)	2.28
Dow Corning 193 Surfactant	2.55
Nacconate 4040 (*Allied*)	110.00

No. 11[1]

RS 410 (*Dow*)	136.50
Refrigerant 11B	55.00
DBTDL (dibutyltindilau-rate)	0.25
TMBDA (tetramethylbutane-diamine)	0.50
Dow Corning 193 Surfactant	2.49
Nacconate 4040 (*Allied*)	113.00

No. 12[1]

PPG 6402 (*Pittsburgh Plate Glass*)	86.0
Refrigerant 11B	38.0
TMBDA (tetramethylbutane-diamine)	1.0
Dow Corning 193 Surfactant	1.0
PPG 6403 (*Pittsburgh Plate Glass*)	100.0

Antifoams*

No. 13[2]

Lecithin	25.0
"Ucon" LB1715	10.0
Polyglycol 400 Dilaurate	8.7
Kerosene	13.0
Mineral Seal Oil	47.8

*See also Chapter 3, Vol. 1.

No. 14[3]

Kerosene	50-70
Stearic Acid	10-15
Polyoxyethylene Tallate	2-6
Alumina Hydrate (fine powder)	5-10
or Magnesium Carbonate	10-30
Water	to suit

References

1. Dow Corning Corp.
2. US Patent 2 727 009
3. US Patent 2 773 041

chapter 7

GASOLINE EMULSIONS

Dissolve the triethanolamine in water and add the mixture of the other ingredients slowly while stirring vigorously. The stability of these emulsions is improved considerably if they are passed through a colloid mill.

No. 1[1]

Triethanolamine	1.5
Water	1.0
Oleic Acid	1.5
Butanol	5.0
Gasoline	45.0

No. 2[1]

Triethanolamine	1
Water	1
Oleic Acid	1
Butanol	5
Gasoline	45

No. 3[1]

Triethanolamine	0.5
Water	1.0
Oleic Acid	0.5
Butanol	5.0
Gasoline	450.

No. 4[1]

Triethanolamine	0.5
Oleic Acid	1.0
Water	1.0
Butanol	5.0
Gasoline	45.0

Alcohol Emulsions

No. 5[1]

Triethanolamine	1
Water	1
Oleic Acid	1
Butanol	2
Alcohol	4

No. 6[1]

Triethanolamine	1
Water	1
Oleic Acid	1
Butanol	4
Alcohol	2

Kerosene–Gasoline Emulsions

No. 7[1]

Triethanolamine	1
Water	1
Oleic Acid	1
Butanol	5
Kerosene	20
Gasoline	25

Alcohol–Gasoline Emulsions

No. 8[1]

Triethanolamine	1
Water	1
Oleic Acid	1
Alcohol	3
Gasoline	45

No. 9[1]

Triethanolamine	1
Water	1
Stearic Acid	5
Alcohol	3
Gasoline	45

No. 10[1]

Triethanolamine	1
Water	1
Stearic Acid	3
Alcohol	3
Gasoline	45

No. 11[1]

Triethanolamine	1
Water	1
Stearic Acid	2
Alcohol	3
Gasoline	45

No. 12[1]

Triethanolamine	1
Water	1
Stearic Acid	3
Alcohol	2
Gasoline	45

No. 13[1]

Triethanolamine	0.5
Water	1.0
Stearic Acid	3.0
Alcohol	3.0
Gasoline	45.0

No. 14[1]

Triethanolamine	0.25
Water	0.50
Stearic Acid	3.00
Alcohol	3.00
Gasoline	45.00

No. 15[1]

Triethanolamine	0.25
Water	0.50
Stearic Acid	2.00
Alcohol	2.00
Gasoline	45.00

No. 16[1]
(0.25% water)

Triethanolamine	1.75
Water	0.85
Stearic Acid	14.00
Alcohol	14.00
Gasoline	315.00

No. 17[1]
(0.5% water)

Triethanolamine	1.75
Water	1.75
Stearic Acid	14.00
Alcohol	14.00
Gasoline	315.00

No. 18[1]
(0.75% water)

Triethanolamine	1.75
Water	2.60
Stearic Acid	14.00
Alcohol	14.00
Gasoline	315.00

No. 19[1]
(1% water)

Triethanolamine	1.75
Water	3.50
Stearic Acid	14.00

Alcohol	14.00
Gasoline	315.00

No. 20[1]
(1% water)

Trihydroxyethylamine Linoleate	21.00
Gasoline	315.00
Butanol	35.00
Water	7.00
Triethanolamine	10.50

No. 21[1]
(5% water)

Trihydroxyethylamine Laurate	21.0
Gasoline	315.0
Butanol	35.0
Water	17.5
Triethanolamine	10.5

No. 22[1]
(10% water)

Trihydroxyethylamine Laurate	35.0
Gasoline	315.0
Butanol	56.0
Water	35.0
Triethanolamine	17.5

No. 23[2]

Gasoline	70.0
Alcohol	9.0
Water	20.3
Bentonite	0.7

No. 24[3]

Gasoline	80.0
Water	19.3
Bentonite	0.7

No. 25[4]
Anti-Knock Fuel

Triethanolamine	18
Oleic Acid	58
Butyl Cellusolve	20
Secondary Hexyl Alcohol	25
Water	75
Gasoline	750

No. 26[5]
Solidified Gasoline

Coconut Oil	50 g
Caustic Soda Solution (10%)	46 cc
Water	120 cc
Benzol	60 cc
Gasoline	5 gal

Mix the coconut oil, caustic soda solution, alcohol, and 20 cc of the water, and heat until completely saponified. Then add the remaining 100 cc of water. Add to this the benzol and stir until completely emulsified. Then add the gasoline in small portions, stirring vigorously with a mechanical stirrer. The resulting jelly-like mass forms a very stable emulsion.

No. 27[6]

Stearic Acid	35
Caustic Soda (30%)	7
Alcohol	500
Gasoline	4500

No. 28[7]

Gasoline	0.5 gal
White Soap (finely shaved)	12.0 oz
Water	1.0 pt
Household Ammonia	5.0 oz

Heat the water, add soap, and mix. When the mixture is cool, add the ammonia; then slowly work in the gasoline to form a semisolid mass.

References

1. H. Bennett, *Chemical Formulary*, vol. 3, p. 107, Chemical Publishing, New York, 1936.
2. Personal Communication.
3. Personal Communication.
4. Bennett, *op. cit.* vol. 5, p. 603, 1941.
5. *Ibid.* vol. 2, p. 237, 1935.
6. Klinkenstein, US. Patent 1 848 568 (1932).
7. Bennett, *op. cit.* vol. 4, p. 15, 1939.

chapter 8

FOOD EMULSIONS

No. 1[1]

Butterscotch Icing

A.	Brown Sugar	8 lb
	Butter	4 lb
	Water	4 pt
	Salt	2 oz
B.	Powdered Sugar	25 lb
	Shortening	5 lb
	Liquid Skimmed Milk	2 pt

Boil A to 245°F. Cream up lightly the ingredients of B and into this creamed mass add the hot butterscotch syrup(A). Whip up to the desired consistency.

No. 2[2]

Chocolate Milk, Stabilized

Sodium Alginate	0.8 lb
Irish Moss	0.7 lb
Sugar	44.0 lb
Cocoa	8.8 lb
Milk	96.0 gal

No. 3[3]

Cocoa Icing

Cocoanut Butter	16
Invert Sugar	20
Water	12
Cocoa	20
Icing Sugar	88
Milk Powder	4.750
Salt	.125
Vanilla	.500

Beat together until smooth and glossy.

Artificial Cream

No. 4[4]

Butter, unsalted	2 lb
Milk	2 pt
Sugar	½ oz
Sodium Alginate	⅙ oz

Warm the milk and butter together to 100°F, add the sugar and sodium alginate, and stir thoroughly until dissolved. Homogenize without the addition of further heat and without allowing the mixture to cool.

No. 5[5]

Butter Fat	19.00
Vegetable Fat	10.00
Milk Powder	7.00
Sugar	0.75
Gelatin	1.00
Borax	0.25
Water	62.00
Flavor	to suit

No. 6[6]

To produce a 32%-fat synthetic whipping cream, use 1 lb 10 oz of emulsifying-type shortening, 2 lb 15 oz of water, 8 oz of dry milk solids, and not over 1.5% fat.

To produce a half-and-half cream made from half uncultured butter and half emulsifying-type shortening, use 1 lb 14 oz of butter, 1 lb 10 oz of emulsifying-type shortening, 1 lb of dry milk solids, not over 1.5% fat, and 5 lb 9 oz of fresh water.

No. 7[7]

Synthetic "Cream" for Confectionery

As an example, a whale oil that has been hydrogenated until its melting point is about 30°C is used. A blend is made from 10 lb of this hardened fat, 5 lb of buttermilk, 1 lb 8 oz of sugar, and 12 oz of glycerin. The resulting emulsion is stirred thoroughly with the introduction of air to form a creamy product.

No. 8[8]

Cheese-, Ice Cream-, and Salad Stabilizer

Locust Bean Gum	65
Irish Moss, powdered	35
Karaya Gum	15

When used in the preparation of cream cheese, the undiluted mixture of the three ingredients above is added when the curds are mixed with the cream in the proportion of about 0.5% by weight on a wet basis. The material is heated to about 165°F, homogenized, and then packed hot.

In ice cream, used diluted with sugar in the preferred proportion of 0.5% on a wet basis, the stabilizer prevents crystallization of ice particles and insures a fine, smooth texture.

No. 9[9]

Dairy-Product Stabilizer

Select Irish Moss (pulverized)	20 lb
Ordinary Irish Moss (pulverized)	20 lb
Glyceryl Monostearate	9 lb
Tri-Sodium Phosphate	8 oz

The two grades of Irish moss, the glyceryl monostearate, and the tri-sodium phosphate are heated with 10 times their weight of water in a mixing kettle. Preferably, the heating is done by direct steam injection, which adds to the moisture content and also serves as an agitator to insure a complete mixture of ingredeients. The mixture is brought to the boiling point and held there for approximately 30 minutes, during which time the steam circulates through the batch, the odor of the Irish moss being carried off in the escaping steam.

The batch is then pumped to an atmospheric double-drum drier or its equivalent, and all but approximately 1-6% of its moisture content is removed.

No. 10[10]

Infants' Milk (synthetic)

Sugar	40
Soya Bean Powder	125
Lactose	30
Peanut Oil	20
Dextrin	20
Egg Yolk, Liquid	50
Calcium Lactate	6
Salt	2

Stir in water before use.

Lemon Oil Emulsions

No. 11[11]

Gum Tragacanth	18 oz
Glycerin	1 pt
Water	4 gal
Lemon Oil	4 qt

Place the gum tragacanth in a kettle similar to a baker's mixing machine. Pour the water over it, stir well, and whip until all the lumps have disappeared. Let stand overnight.

Next morning, squeeze the gum solution through a cheese-cloth bag. Add the glycerin to the homogeneous gum solution. Mix for 5 minutes at slow speed.

Put the machine in second speed and gradually add the lemon oil. After 1 qt of oil has been added, it is advisable to put the machine in third speed, otherwise the oil will not be taken up by the gum solution. When all the oil has been added, stop the machine, scrape the walls of the kettle with a long-bladed spatula, and add 5 oz of benzoate of soda in solution to prevent fermentation. Start the machine again in third speed for about 5-10 minutes.

No. 12[12]

Gum Arabic	13 oz	Terpeneless Oil of Lemon	20 oz

Lemon Oil 20 oz
Glycerin 40 oz
Water to make 10 gal

Mix the gum arabic and glycerin, then mix in the terpeneless oil of lemon and the lemon oil; add water slowly with good stirring. Beat intermittently until homogeneous. Pass through an homogenizer.

No. 13[13]

Transparent Lemon Oil Emulsion

Invert Sugar Syrup (80%)	60 lb
Medium Fine Granulated Sugar	24 lb
Water	16 lb
Gelatin (180 Bloom-Min.)	4 oz

In the preparation of the emulsion, 20 oz of terpeneless oil of lemon and 20 oz of regular lemon oil are mixed with enough of the above solution to produce 10 gal. This mixture is then homogenized and left standing for 18 hours, which allows entrapped air bubbles to rise to the surface. Glass containers are filled from the bottom outlet of the emulsion container and tightly stoppered. The resultant emulsion will be practically transparent.

No. 14[14]

Orange Oil Emulsion

Gelatin	4 oz
Water	16 lb
Cane Sugar	24 lb
Invert Sugar	60 lb
Terpeneless Orange Oil	20 oz
Orange Oil	20 oz

Dissolve the gelatin in the water, add the cane sugar, and heat until dissolved. Add the invert sugar and mix well; homogenize.

No. 15[15]

Orange Emulsion (cloudy)

Gum Arabic (best grade)	1 lb
Water	1 pt
Orange Oil	6 dr

Let the gum arabic and water stand overnight; mix and filter. Add the orange oil. Mix and pass through a colloid mill. Use $\frac{1}{2}$ oz per gallon of syrup.

No. 16[16]

Bitter Almond Oil Emulsion

Gum Arabic	10
Bitter Almond Oil	2
Cane Sugar	8
Water	120

Mix the sugar and gum; add to the water, which is agitated by means of a rapid mixer. Then allow to mix until a uniform emulsion results. Add the oil slowly until completely emulsified.

Butter Substitutes

No. 17[17]

Glycerol Monostearate	85 lb
Coconut Oil (76°F titer)	800 lb
Cottonseed Oil, winterized	125 lb
Cultured Milk	50-50 gal

Melt and mix the glyceryl monostearate, coconut oil, and

cottonseed oil; churn with the milk, temper, work, and salt.

No. 18[18]

Soya Bean Lecithin (60-65%)	100
Glyceryl Distearate	100
Melt together and mix in	
Water, hot	100
Add and mix in	
Water, warm	500

One part of this mixture is added to 100 parts plastic margarine.

No. 19[19]

Water	120.00
Galagum C	1.00
Cottonseed Oil	40.00
Caustic Soda	0.02
Butter Flavor	to suit

Dissolve the caustic soda in the water and strew the Galagum C on the surface; bring to a boil while stirring. Run the cottonseed oil and butter flavor slowly into the mixture, with high-speed intermittent stirring.

Mayonnaise

No. 20[20]

Frozen Egg Yolk	120
Winter Oil (Cottonseed)	330
Spice Mixture*	24
Vinegar†	160
Salt	200
Mustard (ground) No. 1	100

Mix the frozen egg yolk (which must be taken out of the refrigerator the night before making the batch) with a part of the vinegar and the spices. Be sure to keep some vinegar back to be added last or if the batch breaks. Allow the oil to flow in slowly.

* Refer to government definitions and standards for food products.

† Use cider vinegar. Dilute 9 gal of vinegar with 1 gal of water.

No. 21[21]

Frozen Egg Yolk	7.5
Winter Oil (Cottonseed Oil)	33.0
Spice Mixture (as in Formula No. 20)	2.0
Vinegar (as in Formula No. 20)	10.0

No. 22[22]

Cottonseed Salad Oil	70.25
Egg Yolk	10.00
Vinegar (50 grain)	10.00
Water	3.90
Salt	1.45
Sugar	3.50
Mustard	0.80
White Pepper	0.10

The quantity of egg yolk may be reduced to 7.5 lb. The sugar may be left out.

No. 23[23]

Frozen Egg Yolk	4½ lb
Winter-Pressed Cottonseed Oil	75 lb
Salt	13 oz
Mustard No. 2	8 oz
White Vinegar	1 qt
Water	2 qt

No. 24[24]

Egg Yolk (fresh)	12 lb
Sugar	2½ lb

Mustard Flour	1 lb
Salt	1 lb 7 oz
Onion Juice (fresh)	2 oz
Paprika	¾ oz
Pepper (white)	¾ oz
Oil (salad)	13 gal
Vinegar (100 grain, white)	4 qt
Water	2 qt

Put the yolks into the beater bowl. Set the beater at high speed. Beat the yolks until they are well broken up—about one minute.

Add the sugar, mustard flour, salt, paprika, and pepper, and beat until creamed. This will require from 2—5 min., depending on the amount of egg white that adheres to the yolks. The more whites present, the longer will be the time required.

Add the oil slowly at first, and then add at the rate of 1 gal /min, until 9 gal have been added.

Run the beater on slow speed while 1 qt of vinegar and 1 qt of water are added and thoroughly mixed in. Then run beater on high speed and slowly add the remainder of oil at the rate of 1 gal every 2 min. Shift the beater to slow speed and add the remainder of the vinegar and water.

Shut off the beater. Scrape down the sides of the bowl. At slow speed, mix the entire batch for 2 min longer. Long beating of the entire mix tends to make the mayonnaise thinner.

No. 25[25]
(modified)

Salad Oil	20 lb
Egg Yolk	2 lb
100-Grade Pectin (Citrus)	1½ oz
Sugar	4¾ oz
Salt	4¾ oz
Mustard	2½ oz
Concentrated Lemon Juice	3 oz
Water	14 oz
Vinegar	20 oz

No. 26[26]

Egg Yolk	14 lb
Vinegar	10 lb
Cottonseed Oil (prime Summer Yellow)	70 lb
Salt	1½ lb
Sugar	3½ lb
Mustard	¾ lb
Pepper	¼ lb

Mix thoroughly in mixing bowls and run through a colloid mill with a clearance of 0.005 in.

Salad Dressings

No. 27[27]

Cornstarch	5
Water	90
Cottonseed Oil	51
French Mustard	25
Salt	5
Sugar	10
Vinegar (white)	42

Mix the cornstarch with a little cold water and add to the remaining water (hot). Stir and boil until it stiffens; set aside to cool. Put the mustard into a bowl and gradually add the oil, stirring all the time. Add the sugar, salt, vinegar, and then the cooled-down cornstarch.

No. 28[28]

Tapioca Starch	7
Water	30
Corn Oil or Cottonseed Oil	40
Salt	2
Cane Sugar	4
Vinegar (white 5%)	17

No. 29[29]

Whole Eggs	9	
Mazola (Corn Oil)	12.00	pt
No. 2 Buffalo Starch	2.40	lb
Water	2.00	gal
Vinegar	7.00	pt
Cerelose (Corn Sugar)	2.40	lb
Salt	9.60	oz
Mustard	9.60	oz
Paprika	0.96	oz

Formula No. 30[30]

Whole Eggs	11	
Mazola	16.00	pt
No. 2 Buffalo Starch	1.80	lb
Water	11.00	pt
Vinegar	7.00	pt
Cerelose	2.40	lb
Salt	9.60	oz
Mustard	9.60	oz
Red Pepper	0.96	oz

No. 31[31]

Citrus Beverages

Orange Oil or Orange Oil blended with Lemon Oil and Tangerine Oil	8 fl. oz
Brominated Vegetable Oil	6 fl. oz
32°Bé Sugar Syrup	8 fl. oz
Gum Arabic	6 oz
Sunset Yellow color	3 oz
Water	104 fl. oz

These ingredients will make one gallon of concentrate; use 2 oz to gal of syrup.

Place in a graduate of suitable size 8 oz of orange oil or a blend of orange and other oils with 6½ oz of brominated oil. Stir this mixture until it is well blended.

The object of using brominated oil of about 1.250 specific gravity, is to increase the specific gravity of the citrus-oil mixture, which is about .880. The specific gravity of the bottled beverage is approximately .028, and therefore the specific gravity of the brominated citrus-oil mixture should be adjusted to this value.

The brominated oil should be added to the citrus oil in small portions and thoroughly mixed and checked with the hydrometer after each addition, until the desired specific gravity has been obtained. The oils must be at room temperature or 70°F, when taking the specific gravity. The specific gravity of the oil mixture can be quickly determined by a special hydrometer with a single reading.

Place the adjusted oils in a can or some other suitable vessel of about 1-gal capacity. Sprinkle the gum arabic on top of the oil and stir with a stick until all the gum has been mixed with the oil and there are no lumps. **This is very important, otherwise problems may develop later on.**

While stirring, mix the syrup with the oil and gum. The gum will congeal at this point. Keep stirring for a few minutes, then add the water while stirring. This will make a total of 1 qt. Add 1 qt of water continuing to stir. By this time, the oil, gum, and water should have formed a smooth emulsion. To the remaining water (2 qt), add 3 oz of Sunset Yellow color in a separate vessel and when it is all dissolved, add this solution to the oil-gum mixture. Stir for about 10 min and then pass twice through the homogenizer.

Use a high-speed mixer to blend this premix properly. Mixing by hand will not keep the oils together. If no high-speed mixer is available, the emulsion should be stirred while in the homogenizer bowl.

Citrus flavors should be made fresh. Do not make more than 2-3 weeks' supply at one time. (All other flavors may be made as much as five or six months in advance.) Be sure to add 5 gr. benzoate of soda to each gallon if the flavor is not to be used within 3 weeks, and store in a cool place. Freezing will spoil the emulsion.

Note: It is found that a blending of 6 oz of orange oil, 0.5 oz of lemon oil, and 1.5 oz of tangerine oil gives a very fine fruity flavor.

When purchasing brominated oil, find out whether it is heavy or light. Also ask the supplier for instructions on how to use his product.

No. 32[32]

Coffee Cream (all-vegetable, kosher)

Promine-D	2.500
Hydrogenated Vegetable Oil*	12.50
Dextrose	5.000
Centrophil SP	0.100
Viscarin B	0.025
Disodium Phosphate	0.250
Citric Acid (to pH 7.0)	0.075
Sufficient water to bring to 100 g	

Melt the hydrogenated vegetable oil and mix with the Promine, Dextrose, and Viscarin B.

To the water, add the disodium phosphate and citric acid. Heat

* Free fatty acids—0.5 max; Wiley melting point—86-90°F.

he water to 160°F and add the dry mix. Pasteurize at 160°F for 30 nin. Following pasteurization, immediately homogenize at 2000 lb pressure. Cool the product.

Depending on the type of water used, the citric acid may be varied to get the desired pH. It appears that the final pH of the product should be about 6.8; however, more information may be necessary from tests on different types.

Coffee Whiteners

No. 33[33]

Liquid

	Range	*Suggested Levels*
Fat	3.0-18.0	10.0
Protein (i.e. sodium caseinate)	1.0-3.0	2.0
Corn Syrup Solids	1.5-3.0	2.5
Sucrose	1.0-3.0	2.5
Atmos® 150	0.3-0.5	0.4
Carrageenan	0.1-0.2	0.2
Stabilizer Salts (i.e. sodium citrate)	0.1-0.3	0.15
Flavor	*q.s.*	*q.s.*
Color	*q.s.*	*q.s.*
Water	*q.s.* to 100	*q.s.* to 100

The dry ingredients are blended with the fat and liquid ingredients and heated to pasteurizing temperature. (This temperature varies, depending on the particular method used.) Once pasteurized the mix is pumped directly to the homogenizer and homogenized at 2,000—2,500 lb/in.^{2} (1,500—2,000 lb/in.^{2} on the first stage and 500 lb/in.^{2} on the second stage) on a two-stage homogenizer, or 1,500—2,000 lb/in.^{2} on a single-stage homogenizer. The product must be cooled rapidly to 38°F and refrigerated or frozen until use.

No. 34[33]

Spray Dried

Vegetable Fat	35-40
Corn syrup solids (42 D.E.)	55-60
Sodium Caseinate	4.5-5.5
Dipotassium Phosphate	1.2-1.8
Emulsifying Agents	
Atmos® 150, Span® 60, Tween® 60 (60/20/20)	0.3-0.5
or Atmos 150, Tween 65 (75/25)	0.2-0.4
Color	*q.s.*
Flavor	*q.s.*
Anticaking Agent	*q.s.*

To prepare a powdered whitener, an emulsion concentrate is formed (prior to spray drying) by dissolving and/or dispersing the various dry ingredients in just enough water to (1) maintain the solids in solution and (2) impart sufficient fluidity to the concentrate so that it may be pumped. The dissolved solids of the concentrate are usually in the 50-60% range, the higher percentage being strongly recommended.

Once the concentrate has been prepared, it is homogenized in such a manner that the fat particles in the dried emulsion will average about 1 μ in diameter. Under normal circumstances this will require about 2,000—2,500 lb/in.^{2} total pressure on a two-stage homogenizer. Spray dry so that the final product has a moisture content of less than 1%; a particle diameter of 125-150 μ; and entrapped fat globules of 1-3 μ maximum diameter.

Ice-Cream Mixes

No. 35[34]

(14% Butterfat)

Sweet Butter	15 lb
Skim- or Whole-Milk Powder	14 lb
Sugar	13 lb
Powdered Egg Yolk	1½ lb
Gelatin	6 oz
Water	60 lb

Dissolve the milk powder in the water and heat to 110°F. Mix together thoroughly the sugar, dry egg yolk, and gelatin powder and add this mixture to the milk stirring continuously. Heat to 145°F, stir vigorously, pour the mass into a homogenizer bowl and homogenize. Cool immediately by placing the can containing the finished product in a water bath. Let it stand for 24 in order to age, then freeze.

No. 36[34]

19% Cream	60.0
Skim-Milk Powder	4.0
Whole Milk	20.0
Sugar	15.0
Powdered Egg Yolk	1.5
Gelatin	0.5

No. 37[35]

Butter Fat	11.00
Serum Solids	10.00
Sugar	18.00
Natural Dutch Blended Chocolate*	4.50
Emulsifier stabilizer†	0.24

* *Ambrosia Chocolate Co., Milwaukee, Wisconsin.*

† *Vegetable Gums—Stable Stearates.*

Meringue Stabilizers

No. 38[36]

(Hot Process)

A.	Cellulose Gum, CMC-7HSP	15
	Agar Agar	7-15
	Starbake-100, powdered	15
	Dextrose	55-63
B.	Egg Whites. Whip to soft peak.	1 qt
C.	Sugar	2½ lb
	Water	3 lb
	Stabilizer	2-2½ oz

Heat AC to boiling. Slowly stream into egg whites and whip to desired consistency at medium speed.

No. 39[36]

Meringue

Promine-R	15.00
Sodium Bicarbonate	1.25
Trisodium Phosphate• $12H_2O$	1.25
Boiling Water	200.00
Citric Acid	0.63
Premium *Albusoy*	1.85
CMC-7HSP	2.50
Corn Syrup Solids	75.00
Sucrose	225.00

Dry-mix the *Promine*-R, bicarbonate, and phosphate. Sift the remaining dry ingredients together. Pour the boiling water into a heated mixing bowl; with constant stirring add the protein mixture. Whip at high speed until stiff. In a steady slow stream add the remaining ingredients. Whip for 2—3 minutes more. With a pastry tube, squeeze onto a cookie tray that has been powdered with starch. Dry in an oven at 170°F for 3 or more hours until completely dry. Serve with all-vegetable whipped toppings, soy chocolate pudding, or chocolate frozen dessert. Extra powdered sugar may be added as desired to alter the properties of the meringue.

No. 40[37]

Salad Dressing

Starch Paste

Vinegar	15.0
Water	30.0
Sugar	10.0
Tapioca Flour	1.5
Corn Starch	3.5

Cook the mixture at 185—195°F for 10 minutes; allow it to cool thoroughly.

Emulsion Base

Corn Oil	26.00
Whole Egg Yolk	2.50
Water	1.00
Salt	0.50
Mustard	0.25

Mix the egg yolk, water, salt, and mustard. While beating, slowly add the corn oil to form a stiff emulsion.

Blend the 60 lb of cold starch paste with the 30 lb of emulsion base to form the finished salad dressing.

No. 41[38]

Pickle-Flavor Emulsion

Oil of Dill Weed	25.00
Oleoresin Capsicum	1.00
Gum Arabic	1.25
Gum Tragacanth	1.00
Benzoate of Soda	0.10
Water	to 100.00

Mix the oil of dill weed and oleoresin thoroughly with agitation. Disperse the gums into the oil phase with vigorous, high-speed agitation. Slowly add the aqueous phase with continuous agitation. Homogenize in a Manton-Gaulin pressure homogenizer.

References

1. F. FIENE and S. BLUMENTHAL, *Handbook of Food Mfg.* Chemical Publishing, New York, 1938.
2. Green, US Patent 2 097 225 (1937).
3. H. BENNETT, *Chemical Formulary*, vol. 1, p. 54, Chemical Publishing, New York, 1933.
4. *Ibid.* vol. 4, p. 131, 1939.
5. *Ibid.* vol. 1, p. 57, 1933.
6. *Ibid.* vol. 5, p. 145, 1941.
7. *Ibid.* p. 145, 1941.
8. *Ibid.* vol. 3, p. 150, 1936.
9. *Ibid.* vol. 5, p. 146, 1941.
10. *Ibid.* vol. 3, p. 151, 1936.
11. FIENE and BLUMENTHAL, Handbook of Food Mfg. (1938).
12. *Ibid.*
13. BENNETT, *op. cit.* vol. 5, p. 119, 1941.
14. FIENE and BLUMENTHAL, Handbook of Food Mfg. (1938).
15. BENNETT, *op. cit.* vol. 4, p. 45, 1939.
16. Personal Communication.
17. BENNETT, *op. cit.* vol. 5, p. 146, 1941.
18. *Ibid.* p. 146, 1941.
19. *Ibid.* vol. 1, p. 47, 1933.
20. FIENE and BLUMENTHAL, Handbook of Food Mfg. (1938).
21. *Ibid.*
22. *Ibid.*
23. *Ibid.*
24. BENNETT, *op. cit.* vol. 5, p. 147, 1941.
25. *Ibid.* p. 147, 1941.
26. *Ibid.* vol. 4, p. 133, 1939.
27. FIENE and BLUMENTHAL *op. cit.* (1938).
28. *Ibid.*
29. C. A. CROWLEY, ed. *Money Making Formulas*, Chemical Publishing, New York, 1938.
30. *Ibid.*
31. BENNETT, *op. cit.* vol. 12, p. 342-3.
32. *Ibid.* p. 371-2.
33. Atlas Chemical Industries, Inc.
34. BENNETT, *op. cit.* vol. 12, p. 371.
35. E. J. FINNEGAN and J. J. SHEURING, "Consumer Preference for Sugar Levels in Ice Cream and Frozen Desserts," *Food Technology*, 16:113-16 (1962).
36. BENNETT, *op. cit* vol. 12, p. 330-1.

37. *Ibid*. p. 388.
38. *Ibid*. p. 389.

chapter 9

LEATHER AND PAPER TREATMENT EMULSIONS

Fat-Liquors

Formula No. 1[1]

Paraffin Oil	60 lb
Sulfonated Cod Oil	40 lb
Borax	2 lb
Water (at 140°F)	20 gal

No. 2[2]

Fig Soap	20 lb
Water	30 gal
Waterless Moellon	30 gal
Sulfonated Cod Oil	20 lb
Paraffin Oil	30 lb
Water to make	50 gal

No. 3[3]

Soap	1.50
Neatsfoot Oil	1.50
Soda	0.25
Water	97.00

No. 4[4]

Sulfonated Cod Oil	2.0
Stock Solution	2.0
Stock Solution	
Cod Oil	50.0
Degras	50.0
Soap	6.0
Bicarbonate of Soda	3.0
Water	5500.0

No. 5[5]

Sulfonated Neatsfoot Oil	20-40
Spindle Oil	10-20
Neatsfoot Oil	100-80

No. 6, 7, 8[5]

	6	7	8
Sulfonated Neatsfoot Oil	30	20	25
Neatsfoot Oil	40	40	30
Mineral Oil	20	20	20
Soap	10	20	25

5% of any of the formulas above dispersed in water forms an emulsion that is used at a pH of 6.5-8.0.

No. 9[6]

Water (150°F)	1 gal
Sulfonated Cod Oil	2 lb
Raw Cod Oil	1 lb
Light Mineral Oil	2 lb
Tallow Soap, dissolved in 1 gal water	1 lb

No. 10[7]

A. Water	25 gal
B. Sulfonated Castor Oil (50%)	25 lb
C. Cold-Test (20°) Neatsfoot Oil	50 gal
Paraffin Oil (28°)	25 gal

Mix A with B; then add C and mix all ingredients at 30°C.

No. 11[8]

Liquoring Emulsifiable Concentrate

Prime Neatsfoot Oil	49
Sulfonated Oil	43
Ethofat 0/15	8

No. 12[8]

Nonionic Fat-Liquoring Stock Emulsion

Ethofat 0/15 or 0/20	5
Oil	45
Water	50

Dissolve the Ethofat and oil while agitating. Add the water to produce the emulsion.

No. 13[8]

Nonionic Fat-Liquoring Emulsifiable Concentrate

35° Neatsfoot Oil	70
Mineral Oil	20
Ethofat 0/20	10

The ingredients are mixed to form a solution. This mixture can be added directly to the fat-liquoring drum containing the wet leather to form the emulsion.

No. 14[8]

Cationic Fat-Liquoring Concentrate

Sheep Oil or Extra Neatsfoot Oil	94.3
Ethomeen 18/15	5.0
Acetic Acid	0.7

The Ethomeen 18/15 is dissolved in the oil and the acetic acid added to the mixture.

No. 15[9]

Cationic Fat-Liquoring Emulsifiable Concentrate

Ethomeen T/15	3
Arquad 2C	1
Mineral Oil	48
Neatsfoot Oil	48

No. 16[10]

Leather Dressing

Tallow	70.0
Petroleum Jelly	3.5
Diglycol Stearate	13.0
Beeswax	9.0
Rosin	2.0
Water	2.0

Leather Finishes and Polishes

No. 17[11]

A. Olive Oil Bar Soap	5
B. Ammonia	1
C. Beeswax	5
Gray Fatty Carnauba Wax	1
D. Glycerin	1
Water	100

Dissolve A in 50 parts of hot water and add B. Heat AB to

90°C. Melt C and add to AB with constant agitation. When ABC is homogeneous, add the remainder of the water and D. Cool ABCD and package.

Note: A mixture of beeswax and carnauba is used because the beeswax alone is too soft and sticky, whereas the carnauba alone would be difficult to emulsify. Such a product is suitable for application to shoe calf and gives a soft brush-polished finish.

No. 18[12]

A. Castile Soap	15 lb
B. Carnauba Wax (No. 1 Yellow)	5 lb
Candelilla Wax	5 lb
Water	to make 10 gal

Dissolve A in 8 gal of water by boiling for 1 hour, adding water as it is lost by evaporation. Add molten B and continue boiling for 2 more hours. Add enough water to make a total of 10 gal. Strain the product through four layers of cheese cloth and package.

No. 19[13]

A. Gray Fatty Carnauba Wax	20 lb
Turpentine	10 lb
B. Olive Oil Soap	10 lb
Water	16 gal
Turkey Red Oil (50%)	20 gal

Melt A in a steam-jacketed mixing pan. Dissolve B in the hot water and add to A with constant stirring and boil the mixture for 10 min. Turn off the steam and add the Turkey Red Oil. Pass the hot liquid through a water-cooled colloid mill, which cools and emulsifies the product at the same time. Glue may be used as a stabilizing colloid for the emulsion.

No. 20[14]

Gray Fatty Carnauba Wax	20 lb
Turpentine	10 lb
Olive Oil Soap	10 lb
Water	16 gal
Rabbit Skin Glue	40 lb
Water	16 gal

No. 21[15]

Shoemakers' Wax Burnishing Polish

Carnauba Wax	10
Turpentine	10
Duponol WA Paste	10
Iron Chloride or Acetate	3
Acetic Acid (10%)	2
Water	65

Melt the wax and turpentine together and pour into a solution of the Duponol in 45 parts of the water at 95-100°C. Stir vigorously until a smooth emulsion is formed and then stir gently until the emulsion has cooled to approximately 35°C. At this point, add a solution containing the iron salt and acetic acid, dissolved in the remaining 20 parts of water. Continue stirring slowly. The viscosity of the emulsion will be greatly increased at first, after which it will return to its initial value; the resulting emulsion is a smooth, fluid paste.

No. 22[16]

Shellac Wax	3
Montan Wax	3
Carnauba Wax	3
Paraffin Wax	2
Japan Wax	1
Varnish, Acetone	1
Lacquer, Nitrocellulose	1
Lacquer, Acetate	1
Carbon Tetrachloride	20
Water	20
Soda	3
Castor Oil	1
Naphtha	5
Turpentine Substitute	40
Bornyl Acetate	1
Shellac	1

No. 23[17]

Edge-filler for Shoe Factory Use

Soap	15 lb
Yellow Dextrin	5½ lb
Neatsfoot Oil	1½ qt
Oil of Mirbane	1 pt
Gelatin	11½ lb
Formaldehyde	1 qt
Water	1 qt

This is made up with sufficient water to make 60 gal. of solution.

No. 24[18]

Softener for Leather Goods

Glue	3.0
Montan Wax	5.0
Synthetic Wax (Glyceryl Monostearate)	10.0
Sulfonated Neatsfoot Oil	40.0
Glycerin	0.5
Water	41.5

Soften the glue in water and heat the waxes to slightly above melting point; keep at the same temperature while stirring in part of the water. Add the sulfonated oil and keep stirring; add the rest of the water containing the glue and softener.

No. 25[19]

Leather Pasting*

Methylcellulose	89.54
Propylene Glycol	8.97
Casein	1.35
Morpholine	0.14
Water	to suit

* U.S. Patent 2,805,952

No. 26[19]

Fat-Liquoring Leather Treatment

Neatsfoot Oil	10
Span 20	3
Acetylated Lanolin Alcohols	2
Water	to make 100

No. 27[19]

Fur Skin Softener*

Sulfated Sperm Oil (neutralized with ammonia)	90
Carbowax 9000 Disterate	5
Sodium Lauryl Sulfate	2½
Water	2½
Woolfat	20

* British Patent 798,464

No. 28[20]

Degreasing of Pickled Leather

Surfactol 318	2
Surfactol 365	2

Kerosene	40	Water	56

No. 29, 30, 31[21]

Fur Cleaning and Glazing

	29	30	31
Polyethylene Glycol (400) Monolaurate	5	5	5
Polyethylene Glycol (400) Monooleate	3	-	-
Silicone (*Dow Corning* 200/300 cS.)	-	3	-
Acetulan	-	-	3
Perchlorethylene	120	120	120

Dilute 1 part of the above with 7 parts of water; mix to emulsify and then mix with sawdust in tumbling barrel.

No. 32[21]*

Soybean Dimethyl Ammonium Chloride	3 fl oz
Isopropanol	1 fl oz
Poly (Dimethylsiloxane)	1 fl oz
Nonylphenol Polyoxyethylene Ether	1 fl oz
Water	to make 1 gal

Dilute the above with 4 volumes of water and spray on fur and iron at 325°F.

* U.S. Patent 2,861,949.

No. 33[22]

Grease Resistant Paper Coating

Zein G200	100
Oleic Acid	15
Sodium Resinate	20
Water	400
Aqueous Sodium Hydroxide (5%)	<20
Alkali Stable Wax Emulsion (50%)	10
Water	250

Slurry the Zein in cold water below 30°C, then add the sodium resinate and sodium hydroxide. After heating to produce the alkaline dispersion of resin and Zein in water, the oleic acid, wax emulsion, and dilution water are added.

No. 34[23]

Water-Resistant Coating for Paper

Zein	75
Ammonium Resinate	15
Oleic Acid	10
Water	400
Aqueous Ammonia (28%)	15

No. 35[23]

Zein	50
Rosin	40
Oleic Acid	10
Water	300
Aqueous Ammonia (28%)	15
Solvent	20-50

The rosin and oleic acid are first dissolved in the solvent. Zein is slurred in the water with stirring; ammonia is added to the slurry, followed by the cool solvent–oleic acid–rosin solution. Stirring is continued until the solution is complete. The most satisfactory solvents are n-butyl alcohol, ethylene glycol mono-butyl ether, and methyl isobutyl ketone.

References

1. BENNETT, *Chemical Formulary*, vol. 2, p. 288, Chemical Publishing, New York, 1935.
2. *Ibid.* p. 289, 1935.
3. *Ibid.* vol. 4, p. 159, 1939.
4. *Ibid.* p. 159, 1939.
5. *Ibid.* p. 190, 1939.
6. *Ibid.* p. 157, 1939.
7. *Ibid.* vol. 3, p. 200, 1936.
8. *Ibid.* vol. 11, p. 169.
9. *Ibid.* p. 170.
10. *Ibid.* vol. 1, p. 424, 1933.
11. DAVIDSOHN and DAVIDSOHN, *Shoe Creams and Polishing Waxes*, Chemical Publishing, New York, 1938.
12. BENNETT, *op. cit.* vol. 4, p. 178, 1939.
13. DAVIDSOHN and DAVIDSOHN, *op. cit.*
14. *Ibid.*
15. E. I. du Pont de Nemours Co., Wilmington, Del.
16. Veith, British Patent 425 323 (1935).
17. BENNETT, *op. cit.* vol. 3, p. 205, 1936.
18. *Ibid.* vol. 7, p. 219, 1945.
19. *Ibid.* vol. 11, p. 170.
20. Baker Castor Oil Co.
21. BENNETT, *op. cit.* vol. 11, p. 170-1.
22. *Ibid.* p. 237.
23. *Ibid.* p. 239.

chapter 10

MEDICAL EMULSIONS

An emulsion is an ideal formulation for the administration of many topical medicines. The emulsion-form allows uniform application of a small amount of active ingredient on the surface of the skin. Active ingredients can be suspended in an inert, safe vehicle (water). The cooling effect of the evaporation of the water soothes local irritations.

As is true of all formulations in this book, these formulae are presented in good faith by several manufacturers. Proof of the efficacy of any product is, of course, the burden of the manufacturer of the medicine. Therefore, the manufacture of a product based on these formulations should be preceded by satisfaction that the product fulfills the advertised claims. In addition, the suitability for use of any ingredient in a medicine must be established through clinical testing and/or consultation with the government agencies that are involved (in the U.S., the Food and Drug Administration).

Extremely sanitary manufacturing procedures are necessary for any medical product. In addition, the inclusion of fungicides, bacteriacides, and antioxidants in these formulations is highly recommended.

OINTMENT BASES

Oil-in-water Emulsions

Formula No. 1, 2

		No.1[1]	No.2[1]
A.	Amerchol CAB	-----	18.8
	Amerchol H-9	5.0	-----
	Modulan	4.0	-----
	Stearyl Alcohol	1.5	19.6
	Glyceryl Monostearate, s.e.	1.5	-----
	Petrolatum, USP white	37.5	-----
B.	Carbowax 4000*	-----	15.1

* *Union Carbide Chemicals Co.*

Glycerine	-----	10.0
Sodium Lauryl Sulfate	0.5	0.3
Water	50.0	36.2
Perfume and Preservative	*q.s.*	*q.s.*

Add B at 85°C to A at 85°C with moderate stirring. Continue mixing at moderate speed while cooling to room temperature. Remix the following day.

No. 3[2]

A.	Stearyl Alcohol	250.00
	Petrolatum	250.00
B.	Propylene Glycol	120.00
B.	Myrj 52 Polyoxyethylene Stearate (Polyoxyl 40 Stearate, USP)	50.00
	Methylparaben, USP	0.25
	Propylparaben, USP	0.15
	Water	330.00

Heat A and B to 75°C. Add B to A with agitation. Continue agitation until the ointment is set up.

No. 4[3]
Steroidal Emulsion

Steroidal emulsions are reported to release drugs over an extended time period and at predictable rates. The value of lanolin alcohols, which are high in sterol content, in the preparation of ointment bases is indicated by their inclusion in the *British Pharmocopaeia* formula listed below.

Ceralan (lanolin alcohols)	6
Paraffin	24
Petrolatum	10
Mineral Oil	60

No. 5[4]

A.	Amerchol L-101	10.0
	Modulan	3.5
	Glyceryl Monostearate, neut.	13.5
	Spermaceti	1.5
B.	Tween 61*	8.5
	Water	63.0
	Perfume and Preservative	*q.s.*

Add B at 85°C to A at 85°C with moderate stirring. Continue mixing at moderate speed while cooling to room temperature. Remix the following day.

* *Atlas Chemical Industries*

No. 6[5]

A.	Stearic Acid	15.0
	Isopropyl Myristate-Palmitate	1.0
	Arlacel 60 Sorbitan Monosterate	2.0
	Tween 60 Polyoxyethylene Sorbitan Monostearate	1.5
B.	Sorbo Sorbitol Solution, USP	3.0
	Water	77.5
	Preservative	*q.s.*

Add A at 90°C to B at 92°C. Agitate continuously until the emulsion reaches 50°C. Stir intermittently until the product reaches room temperature. Rework and pack on the following day.

No. 7[5]

A.	Stearyl Alcohol	15.00
	Beeswax	8.00
	Arlacel 80 Sorbitan Monooleate	1.25
	Tween 80 Polyoxyethylene Sorbitan Monooleate (Polysorbate 80, USP)	3.75
B.	Sorbo Sorbitol Solution, USP	7.50
	Water	64.50
	Preservative	*q.s.*

Heat A to 70°C, B to 75°C. Add B to A, with stirring. Stir until cool.

No. 8[5]

A.	Mineral Oil	35.0
	Beeswax	17.0
	Lanolin	10.0
	Arlacel 60 Sorbitan Monostearate	2.0
	Tween 60 Polyoxyethylene Sorbitan Monostearate	3.0
B.	Water	33.0
	Preservative	*q.s.*

Heat A to 70°C, B to 72°C. Add B to A slowly, with agitation. Continue agitation until the ointment has cooled.

No. 9[5]

A.	Petrolatum	15.0
	Mineral Oil	15.0
	Spermaceti	5.0
	Arlacel 40 Sorbitan Monopalmitate	5.0
	Tween 40 Polyoxyethylene Sorbitan Monopalmitate	5.0
B.	Water	55.0
	Preservative	*q.s.*

Procedure as in No. 8.

No. 10[5]

A.	Stearyl Alcohol	21.500
	Petrolatum	21.500
B.	Brij 30 Polyoxyethylene Lauryl Ether	3.500
C.	Brij 35 Polyoxyethylene Lauryl Ether	4.500
	Propylene glycol	12.000
	Methylparaben, USP	0.025
	Propylparaben, USP	0.015
	Water	73.000

Heat A to 80°C, then add the Brij 30. Heat C to 80°C and add to the AB mixture. Continue agitation to set point.

No. 11[5]

Cottonseed Oil	85.0
Tween 80 Polyoxyethylene Sorbitan Monooleate (Polysorbate 80, USP)	0.5
Water	14.5
Preservative	*q.s.*

Mix the Tween 80 and water. Add to an equal volume of oil to make a premix emulsion. Pass this emulsion through a colloid mill, adding the remainder of the oil in about five increments in subsequent passes through the mill.

No. 12[5]

A.	Cetyl Alcohol	20.0
	Mineral Oil	20.0

	Arlacel 80 Sorbitan Monooleate	0.5
	Tween 80 Polyoxyethylene Sorbitan Monooeate (Polysorbate 80, USP)	4.5
B.	Water	55.0
	Preservative	*q.s.*

Procedure as in No. 8.

Water-In-Oil Emulsions

No. 13[5]

A.	Mineral Oil	50.0
	Beeswax	10.0
	Lanolin	3.1
	Arlacel 83 Sorbitan Sesquioleate	1.0
B.	Borax	0.7
	Water	35.2
	Preservative	*q.s.*

Heat A to 70°C, B to 72°C. Add B to A slowly, with agitation. Pack at 55-60°C.

No. 14[5]

A.	Petrolatum	54.0
	Arlacel 83 Sorbitan Sesquioleate	6.0
B.	Water	40.0
	Preservative	*q.s.*

Heat A to 65°C. Add B, which has been heated to 67°C, agitating thoroughly during the addition.

No. 15[5]

A.	Petrolatum	40.0
	Mineral Oil	15.0
	Beeswax	4.0
	Arlacel 83 Sorbitan Sesquioleate	6.0
B.	Water	35.0
	Preservative	*q.s.*

Heat A to 70°C, B to 72°C. Add B to A slowly, with agitation. Homogenize when cool.

SPECIAL PURPOSE OINTMENTS, CREAMS, AND LOTIONS

No. 16[6]

Acne Cream

Dehydag® WAX N	15.0
Rice Starch	5.0
Powder Pigment	1.0
Titanium Dioxide	2.0
Resorcin	1.0
Menthol	0.3
Vancide ZP	1.0
Leukichtol	3.0
Precipitated Sulphur	2.0
Water	69.7

Acriflavine Ointments

No. 17[6]

Acriflavine	1.0
Cetiol V	20.0
Emulgade F Special	30.0
Distilled Water	to 100.0

No. 18[6]

Acriflavine	1.0
Amphocerin K	25.0
Peanut Oil	10.0
Distilled Water	to 100.0

No. 19[7]

Aluminum Acetate (Burow's Emulsion) Lotion

A.	Lanolin	10.0 g
	Peanut Oil	40.0 ml
	Tween 80 Polyoxyethylene Sorbitan Monooleate (Polysorbate 80, USP)	2.0 ml
B.	Zinc Oxide	10.0 g

	Talc	10.0 g
C.	Aluminum Acetate Solution	2.0 ml
Water		to 100.0 ml
Preservative		*q.s*

Heat A to 60°C, mix well, and cool to room temperature. Add B, milling if necessary. Then add C, followed by the water.

No. 20[8]

Topical Antibiotic, Anesthetic and Anti-inflammatory

A.	Tegin 515	15.0
	Tegamine P-13	0.7
B.	Water	69.9
	Phosphoric Acid (10% solution)	1.0
C.	Water	10.0
	Neomycin Sulfate	0.5
	Dibucaine Base	0.5
	Phosphoric Acid (10% solution)	1.4
D.	Hydrocortisone Acetate	1.0

Heat A and B to 80°C, slowly and with good agitation add B to A, cool with agitation until the emulsion becomes smooth. Slowly and with good agitation add C, continue cooling with agitation, and mix in D.

No. 21[8]

Topical Anesthetic and Antipruritic

Tegone A	20.00
Tegamine P-13	0.50
Tegosept P	0.10
Water	74.00
Propylene Glycol	2.00
Polyethylene Glycol 1000	2.00
Dibucaine Citrate	1.00
Citric Acid	0.25
Tegosept M	0.15

No. 22[9]

O/W Benzyl Benzoate Ointment

A.	Mineral Oil	5.0
	Stearic Acid	10.0
	Benzyl Benzoate	15.0
	Arlacel 60 Sorbitan Monostearate	7.0
	Tween 60 Polyoxyethylene Sorbitan Monostearate	3.0
B.	Water	60.0
	Preservative	*q.s.*

Slowly add B at 70°C to A at 65°C, with agitation. Continue agitation to 40-45°C. Pour into jars.

No. 23[10]

Burn Cream

Tegacid Special	13.5
Tegone A	5.4
Isopropyl Myristate	8.1
Peach Kernel Oil	8.00
Benzocaine	2.00
Hexachlorophene	0.25
Dilauryl Thiodipropionate	0.05
Dow Corning 200 Fluid	1.00
Water	56.42
Propylene Glycol	5.00
Versene 100	0.03
Sodium Lauryl Sulfate	0.10
Tegosept M	0.15

No. 24[11]

Topical Burn and Wound Medication

Acetulan	1.00
Benzocaine	1.00
Camphor	0.10

Menthol	0.10
Pyrilamine Maleate	0.25
Hexachlorophene	0.02
Oleyl Alcohol	4.00
Dipropylene Glycol	1.00
Genetrons 152a/11* (15:85)	92.53

* *Gen. Chem. Div., Allied Chem. Corp.*

No. 25[12]

O/W Boric Acid Ointment

A.	Stearic Acid	10.0
	Cetyl Alcohol	5.0
	Mineral Oil	25.0
	Myrj 52 Polyoxyethylene Stearate (Polyoxyl 40 Stearate, USP)	5.0
B.	Boric Acid	10.0
	Sorbo Sorbitol Solution, USP	5.0
	Water	40.0
	Preservative	*q.s.*

Heat A to 70°C, B to 72°C. Add B to A slowly, with agitation, stir until cool, and pass through an ointment mill.

No. 26[12]

W/O Calamine Cream

A.	Gel Base	30
	Arlacel 186 — 1 part	
	Sorbo — 9 parts	
B.	Beeswax	1
	Ceresin Wax	1
	Mineral Oil	10
Calamine, USP		17
Water		41
Preservative		*q.s.*

Make the intermediate gel base (A) by slowly adding the Sorbo to the Arlacel 186 with agitation. Blend B into A, heating to 70°C. Add the Calamine to AB and mix until uniformly dispersed. Heat the water to 72°C and add it to the blend. Continue agitation until cooled to room temperature. Milling is suggested for improved stability and shelf-life.

No. 27[13]

Chest Rubbing and Cough Balm

Emulgade® F	18.0
Cetiol® V	5.0
White Mineral Oil	5.0
Vegetable Oil	5.0
Nicotinic Acid Ester	0.5
Turpentine Oil	5.0
Pine Needle Oil	5.0
Oil of Thyme	1.0
Rosemary Oil	1.0
Eucalyptus Oil	1.0
Oil of Juniper	0.5
Oil of Sage	0.5
Camphor	3.0
Water	49.5
Preservative	*q.s.*

No. 28 [13]

Chilblain Ointment

Dehydag® Wax N or SX	16.0
Cetiol® V	4.0
White Petroleum Jelly	12.0
Camphor	4.0
Menthol	0.8
Ethyl Alcohol (96%)	3.2
White Mineral Oil	4.0
Boric Acid	1.0
Water	55.0
Preservative	*q.s.*

No. 29[13]

Cod Liver Oil Ointment

Cod Liver Oil	40.0
Gelatine, white	1.0
Dehydag Wax N or SX	8.0
Distilled Water	to 100.0
Preservative	*q.s.*

No. 30[14]

W/O Emollient Ointment

A.	Acetylated Lanolin	9.0
	Lanolin Alcohols Extract	12.0
	Mineral Oil	20.0
	Microcrystalline wax (m.p. 170-175°F)	12.0
	Arlacel 83 Sorbitan Sesquioleate	2.0
B.	Sorbo Sorbitol Solution, USP	2.0
	Water	43.0
	Preservative	*q.s.*

Heat A to 75°C, B to 75°C. Add B to A with agitation, and stir until the mixture congeals and is completely cooled. Then homogenize the emulsion at a temperature not exceeding 40°C.

No. 31[14]

O/W Emollient Ointment

A.	Stearyl Alcohol	13.0
	Petrolatum	15.0
	Acetylated Lanolin	15.0
	Lanolin Alcohols Extract	5.0
	Arlacel 83 Sorbitan Sesquioleate	2.0
B.	Myrj 52 Polyoxyethylene Stearate (Polyoxyl 40 Stearate, USP)	5.0
	Propylene Glycol	12.0
	Water	33.0
	Preservative	*q.s.*

Heat A to 75°C, B to 75°C. Add B to A with agitation, stirring until the mixture congeals.

No. 32[15]

Anti-Frostbite Ointment

Flowers of Camphor	4.0
Menthol	0.8
Alcohol	3.2
Boric Acid	1.0
Liquid Paraffin	4.0
Cetiol® V	4.0
Dehydag Wax N or SX	16.0
White Petrolatum	12.0
Distilled Water	to 100.0

No. 33[15]

Hemorrhoids Ointment

Dehydag® Wax N or SX	10.0
Cetiol® V	6.0
Lanolin	20.0
White Petroleum Jelly	20.0
Procaine Hydrochloride	1.0
Menthol	0.2
Basic Bismuth Subgallate	5.0
Zinc Oxide	5.0
Rice Starch	5.0
Witch Hazel Extract, dist.	10.0
Water	17.8
Preservative	*q.s.*

No. 34[16]

Solubilized Hexachlorophene

A.	Hexachlorophene	1.0
	Tween 40 or Tween 60	9.0
B.	Water	90.0
	Preservative	*q.s.*

Heat A to 70°C, stir to dissolve the hexachlorophene. Cool

to room temperature, agitate the mixture, and add water slowly.

No. 35[16]

O/W Antihistamine Ointment

A. Stearyl Alcohol 25.0
Petrolatum 25.0
Propylene Glycol 12.0
Tween 80 Polyoxyethylene Sorbitan Monooleate (Polysorbate 80, USP) 5.0

B. Water 31.5
Pyrilamine Maleate 1.5
Preservative *q.s.*

Heat A to 70°C, B to 72°C. Add B to A with agitation. Continue agitation until the ointment is set up.

No. 36[16]

O/W Oxyquinoline Sulfate Ointment

A. Stearyl Alcohol 20.0
Mineral Oil 15.0
Myrj 52 Polyoxyethylene Stearate (Polyoxyl 40 Stearate, USP) 5.0

B. Oxyquinoline Sulfate 1.0
Sorbo Sorbitol Solution, USP 10.0
Water 49.0
Preservative *q.s.*

Heat A to 70°C, B to 72°C. Add B to A slowly with agitation. Stir until cool.

No. 37[16]

O/W Pencillin Ointment

A. Stearic Acid 20.0 g
Span 60 Sorbitan Monostearate 10.4 g
Span 80 Sorbitan monooleate 2.0 g
Tween 20 Polyoxyethylene Sorbitan Monolaurate 5.6 g
Water 138.0 ml

B. Sterile 5% sol. anhyd. Na_2HPO_4 10.0 ml
Sterile 5% sol. anhyd. KH_2PO_4 10.0 ml

C. Sodium Pencicillin, 10,000 units/cc 4.0 ml

Weigh A and autoclave the combination at 17 lb of steam pressure for 20—30 min to melt the ingredients and sterilize them. When nearly cool, agitate the mixture thoroughly to give a smooth cream. Add B, then C. Store the ointment in a refrigerator.

No. 38[16]

O/W Glycol Monosalicylate Ointment

A. Silicone Fluid, 1000 cS 1.0
Atmul 84S Glyceryl Monostearate (non-self-emulsifying) 18.0

B. Myrj 59 Polyoxyethylene Stearate 5.0
Glycol Monosalicylate *q.s.*

C. Sorbo Sorbitol Solution, USP 5.0
Water to make 100.0
Preservative *q.s.*

Heat A to 80°C. Add B to A just prior to emulsification, adjusting the temperature of AB to 70°C. Heat C to 72°C. Add AB C. Stir until set.

No. 39[16]
O/W Methyl Salicylate Ointment

A. Cetyl Alcohol 20.0
Myrj 59 Polyoxyethylene Stearate 5.0
B. Methyl Salicylate 12.0
Menthol 1.0
C. Sorbo Sorbitol Solution, USP 5.0
Water 57.0
Preservative *q.s.*

Heat A to 70°C, B to 70°C, C to 72°C. Add C to B, then add B C to A. Agitate to 40°C and pour into jars. (The ointment will set up in about an hour.)

No. 40[17]
Methyl Salicylate/Menthol Analgesic Balm

Polawax 15.0
Methyl Salicylate 10.0
Menthol 5.0
Mineral Oil, 70 vis. 10.0
Deionized Water 60.0

No. 41[18]
Salicylic Acid Lotion

Salicylic Acid 1.0 g
Isopropyl Palmitate 3.0 g
Isopropyl Alcohol, 99% 70.0 ml
Tween 80 Polyoxyethylene Sorbitan Monooleate (Polysorbate 80, USP) 3.0 g
Water to make 100.0 ml
Preservative *q.s.*

Dissolve the salicylic acid in the alcohol. Add the Tween 80 and isopropyl palmitate, mixing well. Then add the water, agitating the mixture thoroughly.

No. 42[18]
Scalp Lotion

A. Benzoic Acid 4.0 g
Tincture Capsicum 0.1 ml
Undecylenic Acid 5.0 ml
Isopropyl Alcohol 25.0 ml
B. Tween 20 Polyoxyethylene Sorbitan Monolaurate 19.0 ml
C. Benzalkonium Chloride, 12.8% solution 0.8 ml
Water to make 100.0 ml

Mix A well, dissolving the medicaments. Add B and blend, then add C slowly with agitation.

No. 43[18]
O/W Silicone Ointment

A. Stearic Acid 12.5
Arlacel 60 Sorbitan Monostearate 10.0
Tween 60 Polyoxyethylene Sorbitan Monostearate 6.0
Silicone Fluid, 1000 cS 10.0
B. Water 61.5
Preservative *q.s.*

Heat A to 80°C, B to 85°C. Add B to A with agitation. Continue the agitation to 35°C and pack.

No. 44[19]
Skin Abrasion Ointment for Children

Dehydag® Wax N or SX 7.0
Cetiol® V 16.0
Lanolin, anhydrous 10.0
White Petroleum Jelly 13.0
Urea 2.0
Zinc Oxide 5.0

Rice Starch	5.0
Basic Bismuth Subgallate	5.0
Ethyl Alcohol 96%	5.0
Perugen	5.0
Menthol	0.5
Witch Hazel Extract, dist.	10.0
Water	16.5
Preservative	*q.s.*

No. 45[20]

Sulfadiazine Ointment

Tegacid Special	15.00
Petrolatum	5.00
Lanolin	3.00
Cetyl Alcohol	2.00
Water	59.85
Glycerin	10.00
Tegosept M	0.15
Sulfadiazine	5.00

No. 46[21]

W/O Vitamin E Ointment

A.	Petrolatum	28.0
	Vitamin E	*q.s.*
	Arlacel 83 Sorbitan Sesquioleate	2.0
	Tween 80 Polyoxyethylene Sorbitan Monooleate (Polysorbate 80, USP)	1.0
B.	Water to make	100.0
	Preservative	*q.s.*

Heat A to 85°C, B to 87°C. Add B to A slowly with moderate agitation. Stir until cool.

No. 47[22]

Zinc Oxide Ointment

Water	76.0
Carbopol 934	0.8
Sodium hydroxide (10% solution)	3.2
Zinc oxide	20.0

Carefully disperse the Carbopol 934 in the water. After dispersion is complete, add the sodium hydroxide with slow agitation to prevent excessive inclusion of air. Add the zinc oxide in the same manner. Mix to homogeneity.

Note: It is important to add the zinc oxide to the *neutralized* Carbopol 934 mucilage. Zinc oxide added to an unneutralized Carbopol 934 dispersion will immediately flocculate. It is also important that the neutralized mucilage have a minimum viscosity of 15,000 cP (Brookfield 20 rpm) and a minimum pH of 7.0. Final pH after zinc oxide addition must be at least 8.0.

No. 48[23]

Nipple Balm

Dehydag® Wax N or SX	10.0
Cetiol® V	10.0
Glycerine	10.0
Lanolin	12.5
White Petroleum Jelly	12.5
Borax	1.0
Perugen	2.0
Ethyl Alcohol 96%	2.0
Basic Bismuth Subgallate	2.0
Zinc Oxide	2.0
Rice Starch	2.0
Sulfonamide, Water-soluble	2.0
Witch Hazel Extract, dist.	10.0
Water	22.0
Preservative	*q.s.*

References

1. American Cholesterol Products, Inc.
2. Atlas Chemical Industries, Inc.
3. Robinson Wagner Co., Inc.
4. American Cholesterol Products, Inc.
5. Atlas Chemical Industries, Inc.
6. Henkel International GMBH, A. H. Carnes Co. agent
7. Atlas Chemical Industries, Inc.
8. Goldschmidt Chemical Corp.
9. Atlas Chemical Industries, Inc.
10. Goldschmidt Chemical Corp.
11. American Cholesterol Products, Inc.
12. Atlas Chemical Industries, Inc.
13. Henkel International GMBH, A. H. Carnes Co. agent
14. Atlas Chemical Industries, Inc.
15. Henkel International GMBH, A. H. Carnes Co. agent
16. Atlas Chemical Industries, Inc.
17. Croda, Inc.
18. Atlas Chemical Industries, Inc.
19. Henkel International GMBH, A. H. Carnes Co. agent
20. Goldschmidt Chemical Corp.
21. Atlas Chemical Industries, Inc.
22. B. F. Goodrich Chemical Co.
23. Henkel International GMBH, A. H. Carnes Co. agent

References

chapter 11

CUTTING OILS, "SOLUBLE" OILS, MISCIBLE OILS

Formula No. 1[1]

Sodium Corn Oil Soap	14
Water	6
Mineral Oil	64
Water White Rosin	10
Carbitol	2
Diethylene Glycol	4

Aqueous emulsions of this oil are stable, even at 200°F. Stable aqueous emulsions are prepared by using 1-35% of this oil.

The oil is clear and will not become cloudy when cooled to a temperature of 60°F. It will not become covered with a film after standing exposed to the air at a temperature of 80°F over a long period of time or at a temperature of 200°F for one day. It will readily emulsify with water after standing exposed to the air at 200°F for two days.

No. 2[2]

Kerosene	60
Sulfonated Castor Oil	20
Alcohol	1
Sulfonated Coconut Oil (neutralized with Rosin Soap)	20

An emulsion is formed when the concentrate (above) is mixed with water.

No. 3[3]

Oleic Acid	15
Rosin	15
Kerosene	10
Trigamine	20

Melt the oleic acid at 85°C, with high-speed stirring. Add the trigamine and continue stirring. Then turn off the heat and add the kerosene slowly, with stirring. This gives a clear transparent liquid that readily forms an emulsion when added to water.

No. 4[4]

A.	Mineral Oil	4
	Aerosol OT (100%)	1
	Glycostearin	1
Water		100

Heat A to 100-120°C, cool to 80°C, and add slowly, with stirring, to the water (heated to 80°C)

No. 5[5]

Rosin	15
Red Oil	15
Mineral Oil	10
Potassium Hydroxide (45%)	5
Isopropyl Alcohol	10
Wetanol or other wetting agent	1/16

Warm together and mix until uniform.

No. 6[6]

A. Mineral Oil (Spindle Oil)	80
Tall Oil, refined	20
B. Caustic Potash (40° Bé.)	6
C. Methylhexalin	1-2

Saponify A with B; clear with C.

No. 7[7]

Mineral Oil	71.00-74.00
Castor Oil	8.25-9.50
Rapeseed Oil	8.25-9.50
Caustic Potash	8.25-9.50
Soda Ash	0.60-1.25

Mix and add to water as desired.

No. 8[8]

Mineral Oil	75
Naphthenic Acid Sludge	25

No. 9[9]

Mineral Oil	75
Naphthenic Soap	25

No. 10[10]

Mineral Oil	32
Oleic Acid	10
Alcohol	3
Trigamine	5

Mix thoroughly when cold to form a clear transparent oil that is readily emulsified when poured into water with slight stirring.

No. 11[11]

Light Mineral Oil	65
Sulfonated Cod Oil	27
Olein	10
Water	5
Caustic Soda	2
Cresol	1

No. 12[12]

Alcohol, denatured	15
Diglycol Laurate	85
Mineral Oil	400

No. 13[13]

Pine Oil	7.5
Carbitol	2.5
Sulfatate (or other water-soluble Sulfonic-Acid Esters or Salts)	2.0
Water	2.0
Diglycol Laurate	36.0
Mineral Oil	160.0

The sulfatate is dissolved in water and mixed well into the other ingredients. This oil disperses in hard waters without scum formation.

No. 14[14]

Sulfonated Castor Oil	15
Carbitol	5
Diglycol Laurate	80
Mineral Oil	410

No. 15[15]

Sodium Abietate	10
Pine Oil	10

Mineral Oil	40

Mix the sodium abietate and pine oil; then add the mineral oil.

No. 16[15a]

Sodium Abietate	31
Rosin Oil	10
Alcohol, denatured	4
Mineral Oil	159

The mixture of the above oils yields a rich creamy emulsion when mixed with water.

No. 17[16]

Oleic Acid	10
Mineral Oil	32
Trigamine	5
Alcohol	3

No. 18[17]

Rosin	2.0
Petronate	10.0
Diethylene Glycol	2.0
Mineral Oil	84.0

No. 19[18]

Olein or Wool Grease	12.25
Animal or Vegetable Oil	7.50
Mineral Oil	2.45
Caustic Soda Solution	77.80

No. 20[18]

A. Paraffin Oil	48.0 fl oz
B. Red Oil	15.0 fl oz
C. Water	34.0 fl oz
D. Sodium Hydroxide	1.8 oz

Dissolve C in D and mix A and B. Add AB to CD while stirring rapidly.

No. 21[19]

Paraffin Oil (28° Bé)	33
Sulfonated Castor Oil (75%)	33
Sulfonated Red Oil (75%)	33

Mix at 40°C.

No. 22[20]

Paraffin Oil (28 to 30° Bé)	25.0
Rosin	2.2
Oleic Acid	2.2
Caustic Soda	0.3
Water	1.0
Alcohol	0.7

No. 23[21]

Potash Soap	30
Olein	7
Cyclohexanol	2
Paraffin Oil	32

No. 24[22]

Sodium Naphthenic Sulfonate	12.00
Naphthenic Acid	3.00
Sodium Hydroxide	0.50
Alcohol	0.50
Water	2.00
Montan Wax	0.05
Paraffin Oil	81.95

No. 25[22]

Sodium Naphthenic Sulfonate	12.00
Rosin	4.60
Sodium Hydroxide	0.50
Alcohol	0.50
Water	2.00
Montan Wax	0.05
Paraffin Oil	80.35

No. 26[23]

Pine Oil	2.5
Red Oil	12.0
Caustic Potash (50%)	4.0
Alcohol, anhydrous	5.0

Potassium Oleo-Abietate	16.0

No. 27[23a]

Pine Oil	80
Double Sulfonated Castor Oil	20

No. 28[24]

Pine Oil	12
Red Oil	9
Carbitol	8
Diglycol Laurate	65
Sulfatate (or other water-soluble sulfonated hydrocarbon)	6

No. 29[25]

Miscibol	16.0
Oleic Acid	12.0
Caustic Potash (45%)	3.0
Isopropanol	5.0
Pine Oil	2.5

No. 30[26]

Sodium Abietate	25
Oleic Acid	10
Pine Oil	4
Isopropanol	6

No. 31[27]

A. Steam-Distilled Pine Oil	50
B. Sulfonated Castor Oil (75%)	50
C. Caustic Soda (27°Bé)	10
Water	40

Heat A to 38°C in a lead-lined tank, add B, then add C gradually with agitation, maintaining the temperature at 38°C. When a nearly clear solution is obtained, add the water slowly, continuing the agitation, then cool rapidly.

No. 32[28]

A. Soya Oil Fatty Acid or Linseed Oil Fatty Acid	25
B. Pine Oil	35
C. Water	10
D. Caustic Potash (50°Bé)	10
E. Pine Oil	160

Mix A and B and warm gently with C, D, and E.

No. 33[29]

Spindle Oil, refined 4-5 E (20°C)	50.0
Oleic Acid, distilled	20.0
Sodium Hydroxide (29°Bé)	about 12.5
Alcohol, to make clear	about 12.5

No. 34[30]

A. Spindle Oil	100.0
Oleic Acid	12.0
B. Butanol	7.0
Triethanolamine	0.5
Caustic Potash (38°Bé)	4.5

Heat A to about 70°C and add B. Add the caustic potash and allow the mixture to clarify while hot.

No. 35[31]

Tall Oil, refined	20
Spindle Oil, refined	72
Caustic Soda (38°Bé)	8
Alcohol	to clear
Fatty Acid	a little, to clear

No. 36[32]

Naphthenesulfonic Acids	15
Olein (or Liquid Wool Fatty Acid)	5-7
Spindle Oil, refined (60°C)	75
Caustic Potash (25°Bé)	until neutral
Hexalin and Tetralin (1:1)	3-4

No. 37[33]

Rosin Soap	25
Red Oil	10
Pine Oil	4
Alcohol, anhydrous	6

No. 38[34]

Olein	16
Spindle Oil, refined	73
Caustic Soda (36°Bé)	6
Alcohol, denatured	5

No. 39[35]

Olein	20
Spindle Oil, refined	72
Caustic Potash (40°Bé)	8

No. 40[35]

Tall Oil, refined	20
Spindle Oil, refined	70
Caustic Potash (36°Bé)	10

No. 41[35]

Tall Oil, refined	20
Spindle Oil, refined	72
Caustic Soda (36°Bé)	8

No. 42[35]

Train Oil Fatty Acid	20
Spindle Oil, refined	70
Caustic Potash (40°Bé)	7
Hexalin	3

No. 43[35]

Naphthenic or Hydroxy Fatty Acid	12
Spindle Oil, refined	80
Caustic Soda (36°Bé)	6
Benzoline	2

No. 44[35]

Colophony (Rosin)	8
Olein	6
Spindle Oil, refined	78
Caustic Soda (24°Bé)	8

No. 45[35]

Turkey Red Oil	15
Olein	10
Spindle Oil, refined	65
Caustic Potash (40°Bé)	10

No. 46[35]

Bone Fat Fatty Acid	20
Olein	5
Spindle Oil, refined	63
Caustic Soda (24°Bé)	12

No. 47[35]

Lanolin Fatty Acids	16
Spindle Oil, refined	75
Caustic Potash (24°Bé)	9

No. 48[35]

Woolfat Fatty Acids	7
Lanolin Fatty Acids	8
Spindle Oil, refined	77
Caustic Soda (24°Bé)	8

No. 49[35]

Olein	18
Spindle Oil, refined	45
Caustic Potash (40°Bé)	7
Water	30

The amount of caustic must be varied, since the saponifiable materials are not constant in acidity

No. 50[36]

Glyceryl Monostearate	75
Sperm Oil	150
Sodium Salt of Sulfonated Oleyl Alcohol	50
Water	1200

Warm and mix vigorously.

No. 51[37]

Sulfonated Oil	11.5
Soap	17.5
Cottonseed Oil	8.5
Oleic Acid	10.5
Mineral Oil	2.0
Water	50.0

No. 52[38]

Carbitol	5
Diglycol Laurate	95
Orthodichlorbenzol	400

No. 53[39]

Sulfonated Castor Oil	70
Turpentine	10
Tetrahydronaphthalene	15
Pine Needle Oil	5

No. 54[39]

Tetralin	15
Turpentine	10
Pine Needle Oil	5
Sulfonated Vegetable Oil	70

No. 55[40]

Sal Soda	$\frac{1}{4}$ lb
Lard Oil	$\frac{1}{2}$ pt
Soft Soap	$\frac{1}{2}$ pt
Water	to make 10 qt

Boil the above ingredients for 30 min and place in suitable containers.

No. 56[41]

A. Mineral Oil	94
B. Ethomeen S/15	4
Ethomeen S/12	2
Water	*q.s.*

Dissolve B in A. The concentrate (AB) disperses readily in water to form a stable, white emulsion. With different grades of mineral oil, the ratio of Ethomeen S/15 to Ethomeen S/12 may require changing to produce the most satisfactory emulsion.

No. 57[41]

A. Mineral Oil	95.0
B. Ethomeen S/12	2.5
Arquad 2C	2.5
Water	*q.s.*

Dissolve B in A. The emulsifiable concentrate (AB) disperses easily in water, forming a stable emulsion. Metals dipped into an emulsion of this type will pick up an oil coating, which acts as a protective layer against corrosion during storage and shipment.

No. 58[42]

Atlas G-3300	2.5
Atlas G-1086	2.5
Petroleum Sulfonate	12.5
Paraffinic Base Oil	82.5

No. 59[42]

Atlas G-3335	2.0
Petroleum Sulfonate	15.0
Naphthenic Base Oil	83.0

No. 60, 61, 62[43]

	Parts by weight		
	60[43]	61[43]	62[43]
A. Surfactol 318	---	18	12.5
Surfactol 340	7.5	18	12.5
Surfactol 365	42.5	---	---
Dipropylene Glycol	50	54	44.83
Tertiary Nonyl Polysulfide	---	5	---
Petroleum Sulphonate	---	5	5
Dye	---	0.02	---
Masking Agent	---	0.15	---
Water	150	150	600
Hexylene Glycol	---	---	20
Dibenzyl Disulfide	---	---	2
Polyoxyethylene Octadecyl Amine	---	---	2
Phenyl Mercuric Acetate (18% solution)	---	---	1

The base grinding fluid is prepared by mixing A, then forming an emulsion of this nonaqueous mixture in water.

No. 63, 64[44]

	63[44]	64[44]
A. Dowfax 9N4 Surfactant	5	4
Dowfax 9N9 Surfactant	3	2
B. Water	50	50
C. Mineral oil	25	15
D. Sodium Nitrite	2	2
E. Dowicide® A Preservative	1	1

Add A to C and mix B,D, and E. Add BDE to AC with stirring.

Metal-Drawing Compounds

No. 65[45]

Lard Oil, prime	40 lb
Diglycol Laurate	10 lb
Water	50 gal

No. 66[46]

Soap, neutral	10 lb
Diglycol Laurate	7 lb
Lard Oil, prime	13 lb
Mineral Oil	20 lb
Water	200 gal

Warm the first four ingredients together until clear and pour into water slowly; mix well. In certain cases it is desirable to add 3 lb of white lead.

No. 67[47]

Diglycol Stearate	6
Glycowax A	2
Water	56

Naphthenic Miscible Oils

No. 73-75[53]

	Parts by weight		
	73	74	75
A. Potassium Naphthenate from			
Heavy Gas Oil	65	48	---
Light Lube Oil	---	---	37
B. Light Lube Oil	20	35	50
C. Common Solvents			
Group 1: water-soluble			
Diethylene Glycol	5	5	5
Group 2: oil-soluble			
Subgroup A			
Pine Oil	---	10	---
Alpha Terpineol	5	---	5
Subgroup B			
Rosin	5	---	---
D. Water	---	2	3

No. 76-78[53]

	76	77	78
A. Potassium Naphthenate from			
Light Lube Oil	37	37	23
B. Light Lube Oil	45	50	58
C. Common Solvents			
Group 1: water-soluble			
Diethylene Glycol	5	5	5
Group 2: oil-soluble			
Subgroup A			
Pine Oil	10	---	---
Alpha Terpineol	---	---	5
Octyl Alcohol	---	5	---
Subgroup B			
Rosin	---	---	5
D. Water	3	3	4

No. 79-81[53]

	79	80	81
A. Potassium Naphthenate from			
Light Lube Oil	20	30	---
Heavy Lube Oil	---	---	13

B. Light Lube Oil	62	55	69
C. Common Solvents			
Group 1: water-soluble			
Diethylene Glycol	5	5	5
Group 2: oil-soluble			
Subgroup A			
Alpha Terpineol	5	10	5
Subgroup B			
Rosin	---	---	5
Cresylic Acid	5	---	---
D. Water	3	---	3

No. 68[48]

Sodium Alginate	1
Tallow	4
Soap	2
Water	195

No. 69[49]

Tallow	3.0
Rape Seed Oil	1.5
Soap	0.5
Water	95.0

Warm together and emulsify. Maintain a pH of 8-9 when using the product.

No. 70[50]

Soluble Alginate	1
Tallow	4
Soap	2
Water	195

No. 71[51]

Tallow	60 lb
Yellow Soap	15 lb
Water	92 gal

Heat and stir until smooth.

No. 72[52]

Sulfated Higher Fatty Alcohol	3
Tallow	20
Water	20

Formula No. 82[53]

A. Potassium Naphthenate from Heavy Lube Oil	30
B. Light Lube Oil	64
C. Common Solvents	
Group 1: water-soluble	
Ethylene Glycol	3
Group 2: oil-soluble	
Rosin	3

No. 83[54]

Lubricant for Cold-Forming Plastics*

Sodium Salt of Sulfated Olive Oil	20.0
Water to make	100.0

Adjust to pH 7.4

* German Patent 1,041,682

No. 84[55]

Emulsifiable Rustproofing*

Mineral Oil	20
Sodium Petroleum Sulfonate	20
Neutral Degras	40
Butyl Carbitol	50

* U.S. Patent 2,862,823

Triethanolamine 1
Water 14

Emulsify with 1-10 parts of water before use.

No. 85[56]
Mold Release

A. DC-200 Silicone Oil* 200.0
B. Carbopol 934 2.0
C. Dodecylamine 1.0
D. Sodium Hydroxide (10% solution) 6.0
E. Water 200.0

Carefully disperse B in E and mix to a thin, cloudy dispersion without lumps. Add D and then C to BE with mild agitation. The dispersion will gel immediately.

* 1000 centistoke silicone oil—*Dow Corning Co., Midland, Mich.*

Lastly, add A in a slow, steady stream with good mixing. A white, stable emulsion forms immediately.

No. 86[56]

A. Water 50.00
B. Carbopol 941 0.25
C. Sodium Hydroxide (10% solution) 1.40
D. Ethomeen C-25* 0.25
E. Mineral Oil 50.00
F. Graphite† 25.00

Carefully disperse B in A. After complete dispersion, add C and then D. Next, disperse F in E and deaerate. Thoroughly mix EF into ABCD.

* *Armour Chemical Co.*, Chicago Ill.
† No. 5XX, *Superior Graphite Co.*, Chicago, Ill.

References

1. US Patent 1 965 935.
2. US Patent 2 021 448.
3. H. BENNETT, *Chemical Formulary*, vol. 4, p. 207, Chemical Publishing, New York, 1939.
4. Personal Communication.
5. BENNETT *op. cit.* p. 207, 1939.
6. *Ibid.* vol. 3, p. 210, 1936.
7. *Ibid.*
8. *Ibid.* vol. 4, p. 187, 1939.
9. *Ibid.*
10. *Ibid.*
11. *Ibid.* p. 207, 1939.
12. *Ibid.* p. 187, 1939.
13. *Ibid.* p. 95, 1939.
14. *Ibid.*
15. *Ibid.* vol. 1, p. 428, 1933.
15a. *Ibid.*
16. *Ibid.* vol. 4, p. 187, 1939.
17. L. Sonnenborn Sons, New York.
18. ERBS, French Patent 484 382 (1917).

18a. Personal Communication.
19. Personal Communication.
20. BENNETT, *op. cit.* vol. 3, p. 210, 1936.
21. *Ibid.* vol 2, p. 303, 1935.
22. Zimmer, US Patent 2 230 556 (1941).
23. BENNETT, *op. cit.* vol. 4, p. 96, 1939.
23a. Personal Communication.
24. BENNETT, *op. cit.* vol. 4, p. 95, 1939.
25. Personal Communication.
26. Personal Communication.
27. Personal Communication.
28. BENNETT, *op. cit.* vol. 3, p. 325, 1936.
29. *Ibid.* vol. 4, p. 187, 1939.
30. *Ibid.*
31. *Ibid.* p. 207, 1939.
32. *Ibid.* vol. 3, p. 213, 1936.
33. *Ibid.* vol. 4, p. 95, 1939.
34. *Ibid.* p. 206, 1939.
35. *Ibid.*
36. *Ibid.* p. 207, 1939.
37. *Ibid.*
38. *Ibid.* p. 95, 1939.
39. Muncheburg, French Patent 674 816 (1929).
40. BENNETT, *op. cit.* vol. 2, p. 303, 1935.
41. *Ibid.* vol. 11, p. 172.
42. *Ibid.* p. 414.
43. Baker Castor Oil Co.
44. The Dow Chemical Co.
45. BENNETT, *op. cit.* vol. 4, p. 202, 1939.
46. *Ibid.*
47. *Ibid.*
48. *Ibid.* vol. 3, p. 211, 1936.
49. *Ibid.* vol. 4, p. 202, 1939.
50. *Ibid.* vol. 2, p. 300, 1935.
51. *Ibid.* vol. 3, p. 209, 1939.
52. WILLIAMS, US Patent 2 152 396 (1939).
53. BRADLEY, US Patent 2 289 536 (1942).
54. BENNETT, *op. cit.* vol. 12, p. 414.
55. *Ibid.* vol. 11, p. 187.
56. Greenwood Textile Supply Co.

chapter 12

PAINT

An emulsified paint possesses many advantages over a solubilized paint. A much higher-molecular-weight polymer can be applied from an emulsion system than from a solvent system of the same viscosity. Thus, the desired high-polymer paint can be created—and endblocked—in the laboratory or in a production plant in an emulsion-paint system. On the other hand, the solvent-based paint contains a lower-molecular-weight polymer that must be oxidized to the desired high polymer after application. Unfortunately, it is hard to stop an oxidizing paint film from cross-linking to a greater extent than is desired. Thus the emulsion paint "dries" faster and has less tendency to become brittle upon aging.

The emulsion paint also possesses several application advantages over the solvent paint. The emulsion paint can be applied to wet or damp surfaces, whereas the solvent paint system requires a dry surface. The emulsion system can be made more highly thixotropic, and therefore easier to handle, than conventional solvent-based systems. The more thixotropic emulsified paint clings to the brush, drips less, and is easier to apply than the solvent system. Because the emulsion paint is water soluble, it is also easier to "clean up." Paint brushes and other tools can be cleaned more easily in water (usually commonly available) than in solvent (less readily available and usually a fire hazard).

Based upon these advantages, latex and emulsified paints have increased greatly in popularity in the last few years.

Formula No. 1[1]

Hydrated Lime	43.0
Hydraulic Cement	19.5
Talc	12.0
Metronite	11.5
Salt	6.5
Mica	5.0
Gum Arabic	1.6
Gum Karaya	0.5
Irish Moss	0.1

Calcium Stearate	0.3

Add water to suit before use.

No. 2[2]

A.	Potato Starch	10
	Water	30
	Caustic Soda (10°Bé)	10
B.	Casein	6
	Water	20
	Caustic Soda (10°Bé)	10
C.	Water	150
	Linseed Oil Varnish	80

Blend A, B, and C separately. Stir these three mixtures together to form a smooth emulsion that will remain stable for 6-8 months.

No. 3[3]

A.	Casein	30.0
	Water	150.0
	Borax	3.5
	Phenol	1.0
B.	Formaldehyde (40%)	2.0
	Water	5.0
C.	Rosin	15.0
	Linseed Oil Varnish	15.0
	White Spirit	10.0

Soak the casein in hot water, then add the borax and the phenol. After 24 hr, mix in B followed by C (which has been previously mixed hot). Since the pigments are more easily wetted in oil than in water, it is advisable to grind them with the oil binder C before this is emulsified with the casein solution. The amount of pigment averages about 6 to 8 times the amount of binder. Casein paints thus made excel by their good brushability and increased water resistance.

No. 4[4]

Dehydrated Castor Oil	33.41
Lithopone	39.81
Casein	3.31
Phenol	0.06
Ammonium Hydroxide	23.41

Mix well until uniform; dilute before using.

No. 5[5]

A.	Shellac	30.0
	Ammonium Bisulfate	4.5
	Water	65.5
B.	Linseed Oil	30.0
C.	Lithopone	380.0

Mix A until dissolved, add B, emulsify AB with agitation, and add C.

No. 6[6]

Water-ground Mica	2 lb
Lithopone or Titanox	6 lb
Casein (80 mesh)	1 lb
Sodium Fluoride	2 oz

No. 7[7]

A.	Montan Wax, double-bleached	7.0
B.	Potassium Oleate	3.0
	Potash, caustic	0.8
	Water-soluble Dye	2.0
C.	Water	87.2

Dissolve B in the boiling water. The boiling solution BC is then slowly poured in small portions into the wax melt. Care must be taken that a drop in temperature does not cause a separation. It is essential that the mixture be stirred constantly while cooling.

No. 8[8]

A.	Crude Wool Fat	36.0
	Refined Tung Oil	4.0
	Rosin	25.0
B.	Ammonia Solution	4.3
	Alcohol	10.0
	Water	120.7

Heat A and B; when A is melted and both A and B are at the same temperature, add A to B and agitate until cold. The product—a viscous, yellowish-white emulsion—may be directly employed as a paint.

No. 9[8]

A quick-drying paint comprises 1000 parts of the emulsion, Formula No. 8, 80 parts chrome oxide, 150 parts titanium white, and 15 parts of a 33% solution of a cobalt–lead–manganese drier.

No. 10[9]

Double-Boiled Oil, with Driers	50
Water	45
Sodium Silicate	5
Pigment	50

The oil, which may be diluted if required with 120°F flash white spirit, should be added to the aqueous phase in a slow stream with rapid and vigorous stirring.

No. 11[10]

A.	Sodium Silicate	10.0
	Ammonia	10.0
	Water	10.0
	Zinc Oxide	5.0
	Sulfur	3.0
	Zinc Dimethyldithiocarbamate	0.5
B.	60% Latex	100.0
	Whiting	200.0
	Spindle Oil	60.0
	Glue	5.0

Blend A and B separately and mix together. The more alkaline varieties of sodium silicate cause precipitation of latex by hydrolysis. If, however, ammonia is added to the solution, this increases the hydroxyl-ion concentration and prevents splitting of the silicate, so that the latex is thickened and rendered stable.

No. 12[11]

A.	Potato Starch	10
	Cold Water	30
	Caustic Soda (10°Bé)	10

Mix the starch with cold water and add the caustic slowly in a thin stream until a transparent thick liquid is obtained.

B.	90-mesh Lactic Casein	6
	Water	20
	Caustic Soda (20°Bé)	10

Soak the casein in the warm water, not over 130°F, and add the caustic while stirring.

C.	Medium Congo Copal	20.0
	Linseed Oil	50.0
	White Spirit	30.0
	Manganese (as resinate)	0.1
	Linseed Oil Varnish	80.0
	Water	150.0

Grind in the required amount

of pigment with the oil varnish and then stir in the water. Run the three solutions together through a homogenizer and the resulting emulsion will be stable for a year. If for export to a hot country, it is advisable to add a little preservative, e.g., metachlorparacresol.

No. 13[12]

Trihydroxyethylamine Linoleate	0.6
Glue	10.0
Water	32.0
Varnish	16.0
Naphtha	4.0
Sodium Ortho Phenyl Phenate	0.1

No. 14[13]

Beeswax	70
Japan Wax	44
Montan Wax	25
Shellac	10
Potassium Carbonate	30
Water	730

Boil together and mix well until emulsified. Cool and add with stirring:

Alcohol	40
Trioxymethylene	50

No. 15[14]

Lithopone	60.00 lb
Asbestine	20.00 lb
Mica Vehicle	8.00 lb
Water	51.00 lb
Dowicide B	0.45 lb
Sodium Alginate	1.00 lb
Raw Chinawood Oil	9.00 lb
Cobalt Drier (6%)	0.48 oz
Wetting Agent	1.68 oz

Place the water (preferably from the hot water tap at approximately 100°F) in the mixer. Add the Dowicide B and stir until dissolved. Add the sodium alginate and stir for 20 min or until completely dispersed. Add the wetting agent and then the chinawood oil (with the drier added to the oil before adding to the water mixture). Stir the oil and alginate mixture for 10 min and then add pigment and filler. Stir for one hour and package.

No. 16[15]

Sodium Silicate	2.00 lb
Soap	0.50 lb
Pale Boiled Linseed Oil	0.25 lb
Water, boiling	1.00 gal

Mix very thoroughly. A little alum may be added to increase water resistance.

No. 17[16]

A. Oiticica Oil	120
Lead Oxide	6
Manganese Dioxide	2

Heat to 250°C, cool to 200°C, and add:

Potassium Silicate	13
Milk of Lime	16
Water	*q.s.*

Agitate violently until cool.

No. 18[17]

Resin Oil Emulsion

A. Ester Gum	62.00
Bodied Linseed Oil	50.00
Pine Oil	21.00

Mineral Spirits	62.00
Cobalt Linoleate Drier (6% Co)	1.25
Lead Linoleate Drier (16% Pb)	4.25
Oleic Acid	24.00
B. Powdered Casein	27.00
Preservative (Moldex)	0.375
Emulsifier (2-amino-2-methyl-1-propanol)	15.00
Water	536.00

Mix the resin, and cut in the mineral spirits with the bodied linseed oil, pine oil, driers, and oleic acid. Place the casein in a mixer, add the water slowly, stirring so that the casein is wetted, add the preservative and emulsifying agent and continue stirring for several minutes.

Add solution A slowly to solution B with constant high-speed agitation until emulsification is complete.

Semipaste Paints

No. 19[18]

Albalith-335, 332, or 836	79.0
Ester Gum Oil Emulsion	62.2

No. 20[18]

Cryptone ZS-830	37.6
China Clay	23.0
Ester Gum Oil Emulsion	62.5

These semipastes are designed to be thinned 2 to 1 with water for brush application.

The illustrative formulae of the emulsion and semipastes given here can be modified in many ways, but in making substitutions, or changing the percentages of the ingredients, the new system must be balanced or breaking of the emulsion may follow.

No. 21[19]

"Piccolyte" Oil Emulsion (V-338)

A. Piccolyte S-100 Resin	27.2
Lithographic Oil No. 8	22.0
Pine Oil	9.3
Mineral Spirits	27.2
Cobalt Linoleate Drier (6% Co)	0.5
Lead Linoleate Drier (16% Pb)	1.9
Oleic Acid	11.9
B. Casein B-5	5.2
Moldex or Dowicide A	0.1
2-amino-2-methyl-1-propanol	2.9
Water	91.8

Dissolve the Piccolyte resin in the mineral spirits and mix with the lithographic oil No. 8, pine oil, driers, and oleic acid.

Place the casein and preservative in a pony or other type of mixer, add the water slowly, stirring so that the casein and preservative are wetted. Add the 2-amino-2-methyl-1-propanol and continue the stirring for several minutes.

Add solution A slowly to solution B with constant agitation

until emulsification is complete.

No. 22[20]

"Piccolyte Oil Emulsion Semipaste"

Pigment (54.5%)

Cryptone ZS-830	62
China Clay	38

Vehicle (45.5%)

Piccolyte Emulsion V-338	93.1
Water	6.9

The pigment, emulsion, and water should be mixed together and then milled on a roller or buhrstone mill at a loose setting.

Other Emulsion Paints

No. 23[21]

(Interior White)

Paint Base

(*Part I*)

Dibutyl Phthalate	22.0
Hexylene Glycol	16.5
Ethylene Glycol	27.5
Tergitol NPX	2.2
Aerosol OT (75%)	3.0
Water	16.5

(*Part II*)

Gelva Emulsion TS-30	267.0

Pigment Slurry

Water	131.0
Tamol 731 (10%)	3.0
Methocel HG 4000 cP (2%)	200.0
Titanox RA50	220.0
Wollastonite Pl	41.0
Gold Bond "R" Silica	39.0

No. 24[21]

Paint Base

(*Part I*)

Hexylene Glycol	15.0
Ethylene Glycol	25.0
Tergitol NPX	2.0
Aerosol OT (75%)	2.7
Water	15.0

(*Part II*)

Gelva Emulsion TS-70	285.0

Pigment Slurry

Water	189.0
Tamol 731 (10%)	10.0
Methocel 4000 cP (2%)	180.0
Titanox RA 50	197.0
ASP 400	88.0
Gold Bond "R" Silica	88.0

No. 25[22]

Flat Wall Finish, White or Light Tint Base

A.	Water	153
	Potassium Tripolyphosphate	1
	Advawet No. 33	4
	Ti-Pure R-610	150
	Alsilate W	100
	Gold Bond "R" Silica	100
	Micromite	50
	Methocel 400 cP (2% solution)	100
	Foamicide 581-B	1
B.	Resyn 12K51	295
	Methocel 400 cP (2% solution)	40
	Water	90
	Hexylene Glycol	15
	Ethylene Glycol	30

Charge A into a mixer with agitation, disperse in a Morehouse or other high-speed mill, then slowly add B.

Interior Latex Paints

No. 26[23]
(Deep Green)

A. Pigment Green B 20.0
ASP 400 Clay 100.0
Celite 281 40.0
Titanox RA-50 75.0
Water 250.0
R and R 551 Lecithin 5.0
Blancol Solution, 20% in water 12.5

B. EVT 50 (Cargill Modified Latex) 434.0
24% Lead Naphthenate 4.0
6% Cobalt Naphthenate 2.0
Water 33.0
Polyco 530 11.0

Grind A for 2—4 hr in a pebble mill and add B.

No. 27[23]
(Deep Gray)

Rutile Titanium Dioxide 214.2
Lithopone 65.3
Superjet Lampblack 1.4
Magnesium Silicate 36.1
Celite 281 42.0
Potassium Tripolyphosphate 1.0
Soya Lecithin, water dispersible 5.3
Water 233.0
15% Alpha Protein Solution 73.5
Dowicides 1 and 7 (1:1), 20% in Carbitol 30.0
Pine Oil 2.5
Sindar No. 19 0.5
Pliolite Latex 165 (48%) 374.0
Nalco 211 (50%) 3.3
Potassium Polyacrylate (15%) 15.0

No. 28[23]
(Tint Base)

Rutile Titanium Dioxide 200.0
Lithopone 70.0
Magnesium Silicate 27.0
Celite 281 33.0
Potassium Tripolyphosphate 1.0
Soya Lecithin, water dispersible 3.4
Dowicides 1 and 7 (1:1), 20% in Carbitol 29.0
Water 181.0
15% Alpha Protein Solution 88.0
Pine Oil 2.1
Sindar No. 19 0.5
Pliolite Latex 165 (48% N.V.) 425.0
Nalco 211 (50%) 4.0
Potassium Polyacrylate (15%) 12.5

No. 29[23]
(Interior Flat Wall)

Titanox RA-50 240.0
Atomite 71.0
Celite 281 52.0
Tetrapotassium Pryophosphate 1.2
15% Casein, KOH cut 118.5
Dylex K-34 Latex (48% solids) 372.0
Water 252.0
Butrol 3.3

Defoamer ED 1.3

Polyvinyl Acetate Emulsion Paints

No. 30[24]

(White Flat Ceiling)

R and R 551 Lecithin	5
Butrol	1
Tergitol NP-14	4
Ethylene Glycol	18
Carbitol	17
Celluflex 179C	20
Titanox RA-50	200
Celite 110	200
Nytal 300	150
Water	133

Add in the order listed and mix as a heavy paste for a minimum of 20 minutes; then add:

Foamicide 581B	2
Water	*q.s.*

Disperse on a three-roll mill and let down with:

2% Methocel 4000	230
Celanese CL-102	234
Water and/or 2% Methocel 4000 (for viscosity adjustment)	85

No. 31[24]

(Interior White)

Elvacet 81-900 Polyvinyl Acetate	287
Dibutyl Phthalate	32
Pigment Grind	
Water	140
Tamol 731, 10% solution	18
Emulphor El-719	2
Polyglycol P-1200	2
Carbitol	35
2% Methocel 4000	50
Ti-Pure R-510 Titanium Dioxide	200
HGO-55 Talc	75
Celite 281	75
Reduction	
Water	56
2% Methocel 4000	130
Dowicide A, 20% in water	5

No. 32[24]

(White Interior Wall)

3% Methocel 4000	80.0
Polyglycol P-1200	2.0
Water	110.0
Titanox RA-50	225.0
Mica C-3000	50.0
Celite 281	25.0
Water	125.0
Water (mill wash)	31.0
Emulphor EL-719	2.2
Aerosol OT-B	2.6
Foamicide 581B	0.8
Carbitol Acetate	20.0
20% Dowicide A	11.5
Everflex G (51%)	353.0
Viscosity Adjustment	
Water and/or 3% Methocel 4000	40.0

No. 33[24]

(Dark Green Interior Wall)

3% Methocel 4000	60.0
Polyglycol P-1200	2.0
Water	208.0
Igepal CTA-639	2.0
Aerosol OT-B	2.6
Butrol	2.5
Titanox RA-50	50.0
Phthalocyanine Green	20.0
Mapico Yellow Dark Orange	35.0

Mica, 325 mesh	50.0
Celite 281	75.0
Carbitol Acetate	20.0
Everflex G (51%)	440.0
Viscosity Adjustment	
Water and/or 3% Methocel 4000	40.0

No. 34[24]

(Universal Tint Base)

A.	2% Methocel 4000	67.0
	Polyglycol P-1200	2.0
	25% Tamol 731	5.0
	Water	25.0
	Titanox RA-50	200.0
	Wollastonite P-1	100.0
	Nytal 300	20.4
	Celite 281	25.0
	Phenyl Mercuric Borate	0.2
	Water	47.6
B.	Water (mill wash)	10.0
	2% Methocel 4000	151.5
	Water	15.0
	Carbitol Acetate	15.0
	Nonic 260	8.0
	Hallco C-451	4.0
C.	Everflex G (51% N.V.)	410.0

Combine the ingredients of A in a change can mixer. Mix to a smooth homogeneous paste, then add the remaining water and pass once through a high-speed stone mill. Premix B, add B to A, and add C with good agitation.

No. 35[24]

(Light Tint Base)

A.	Water	232.0
	2% Methocel 4000	125.0
	Potassium Tripolyphosphate	1.0
	Nopco JMK-1	4.0
	Dowicide A	2.0
	Brij 30	5.0
	Rutile Titanium Dioxide, non-chalking	200.0
	Atomite	70.0
B.	Celite 281	17.5
	Poly-Tex 600	362.0
	Ethylene Glycol	20.0
	Brij 30	2.0
	Atlas G-3300	13.0
	Water	6.0
	Carbitol Acetate	15.0

Grind A for 4—6 hr. Rinse mill and add B.

No. 36[24]

(Deep Tint Base)

A.	Water	203.0
	2% Methocel 4000	83.5
	Potassium Tripolyphosphate	1.0
	Nopco JMK-1	4.0
	Dowicide A	2.0
	Rutile Titanium Dioxide, non-chalking	100.0
	Atomite	100.0
	Celite 281	100.0
	Brij 30	5.0
B.	Poly-Tex 600	362.0
	2% Methocel 4000	74.2
	Ethylene Glycol	20.0
	Carbitol Acetate	15.0
	Brij 30	2.0
	G-3300	12.5

Grind A for 4—6 hr in a pebble

mill. Rinse mill and add B.

No. 37[24]
(Interior–Exterior White)

3% Methocel 4000	60.0
Polyglycol P-1200	2.0
25% Tamol 731	8.0
Phenyl Mercuric Borate	0.2
Water	150.0
Titanox RA-50	200.0
Mica C-3000	25.0
Celite 281	25.0
Talc	135.0
Carbitol Acetate	15.0
Ethylene Glycol	40.0
Water	35.6
Emulphor EL-719	2.0
3% Methocel 4000	80.0
Everflex G (51%)	350.0

No. 38[24]
(Texture Paint)

A.	Water	80.0
	Potassium Tripolyphosphate	1.0
	Advawet No. 33	2.0
	Titanox RANC	87.0
	Hydrite Flat	93.0
	Ethylene Glycol	25.0
	Camel Carb	60.0
	2% Methocel 4000*	75.0
	Foamicide 581-B	1.5
B.	No. 873 Quartz (stir in)	245.0
	Resyn 12K-51	221.0
	Celite 281 (stir in)	42.0
	Water	147.0
	2% Methocel 4000*	85.0
	Carbitol Acetate	5.5
	Perlite No. 600 (stir in)	53.5

* See Formula No. 39 note.

Charge A into a mixer with agitation. Disperse in a Morehouse or other high-speed mill, then slowly add B.

No. 39[24]
(Stipple Paint)

A.	Water	175.0
	Potassium Tripolyphosphate	1.0
	Advawet No. 33	4.0
	Titanox RANC	125.0
	Asbestine 3X	115.0
	Bentonite	10.0
	2% Methocel 4000*	100.0
B.	Resyn 12K-51 (55% N.V.)	290.0
	Water	25.0
C.	Celite 281	180.0
	2% Methocel 4000*	95.0
	Hexylene Glycol	15.0

Grind A, add B, and stir in C.

* Protect the Methocel solution with $\frac{1}{4}$% Dowicide A based on total weight of solution.

No. 40[25]
Interior White Emulsion Paint

Paint Base

A.	Dibutyl Phthalate	19.4
	Hexylene Glycol	15.0
	Neville R7 (108-112°C)	18.6
	Tergitol NPX	2.0
	Aerosol OT	2.0
B.	Methocel 4000 cP (2%)	55.0
C.	Water	134.0
D.	Gelva Emulsion TS 22	211.0

Pigment Slurry			
Water	155	Wollastonite PL	100
Tamol 731 (1%)	20	Gold Bond "R" Silica	50
Methocel 4000 cP (2%)	100	ASP 400	50
Titanox RA 50	135	Albalith 14	45

No. 41[26]

Interior Flat Tint Base

3% Methocel Solution (65HG, 4000DG) (*Dow Chemical Co.*)	100.0
Polyglycol P-1200 (*Dow Chemical Co.*)	2.0
Water	100.0
PMA-18 (*Advance Solvents and Chemicals Corp.*)	1.0
Surfactol 365 (*Baker Castor Oil Co.*)	3.0
Aerosol OT-B (*American Cyanamid Co.*)	2.5
Titanox RA-50 (*Titanium Pigments Corp.*)	200.0
Mica 325 mesh (*Southern Mica Corp.*)	50.0
Celite 281 (*Johns-Manville Corp.*)	25.0
Camel Carb (*H.T. Campbell Sons, Corp.*)	150.0
Foamicide 581-B (*Wyandotte Chemicals Corp*)	0.8
3% Methocel Solution (65HG, 4000DG) (*Dow Chemical Co.*)	50.0
Carbitol Acetate (*Union Carbide Chemicals Co.*)	20.0
Water	165.0
Everflex BG (*Dewey and Almy Chemical Division*)	275.0
	1144.3

No. 42[27]

Semigloss Latex Coating

Ucar Latex 180 (*Union Carbide Chemicals Co.*)	541.03
Titanox RA-45 (*Titanium Pigments Corp.*)	167.50
ASP-100 (*Minerals and Chemicals Philipp Corp.*)	60.30
Water	93.80
25% Tamol 731 Solution (*Rohm and Haas Co.*)	7.00
Bubble Breaker No. 746 (*Balab Inc.*)	5.36
Safflower Oil	26.80
Surfactol 318 (*Baker Castor Oil Co.*)	2.35
Tetronic 504 (*Wyandotte Chemicals Corp.*)	2.35
Ethylene Glycol	29.48
Carbitol solvent (*Union Carbide Chemicals Co.*)	21.11
6% Cobalt Drier	0.11
24% Lead Drier	0.03
Cellosize Hydroxyethyl Cellulose (*Union Carbide Chemicals Co.*)	82.08
QP-15000 solution (2% n.v.)	1039.30

No. 43[28]
Polyvinyl Acetate Latex, Primer-Sealer, Polyvinyl Acetate Base

White Pigment-Extender Dispersion

Vicatmide Solution (25% solids)	5.12
Defoamer	6.40
Ethylene Glycol	15.99
Ti-Pure R-510	58.62
Barytes No. 1 White	74.63
Mineralite 3X	53.31
No. 400 Silver Bond B Silica	53.31

Stir together well for 5-10 min then add

No. 7 Guar Gum Solution (2% solids)	106.61

Pass once through a colloid mill; then add

B-3 PVAc Latex Vehicle	223.47

Then add

No. 7 Guar Gum Solution (2% solids)	319.83

No. 44[29]
Housepaint Primer

Pigments

Rutile Titanium Dioxide	125.0
Basic Lead Silicate	350.0
Magnesium Silicate	250.0
Linaqua	300.0
Kelecin 1081	14.0
Acetic Acid	14.0
Cupric Acetate or Sodium Persulfate	14.0

Driers

6% Cobalt Naphthenate	2.5
6% Manganese Naphthenate	1.0
24% Lead Naphthenate	6.0
Water	325.0

No. 45[30]
Chlorinated Rubber Exterior Paint

A.	Chromium Oxide Green	107.5
	Wet Ground Mica	107.5
	Celite 110	70.7
B.	PE-16 Clear	681.0
	Water	56.2

PE-16 Clear is an emulsion of chlorinated natural rubber. In the preparation of paints, it should not be milled or overheated. The pigments (A) should be dispersed in water and the dispersion then let down with the emulsion vehicle B.

Preparation of PE-16 Clear

1. Into a heavy-duty premixer charge 100 lb of water, 14 lb of 100% acetic acid, 14 lb of cupric acetate (or sodium persulfate), and 14 lb of Kelecin 1081. Mix until crystalline salts are dissolved (5 min).
2. Add 100 lb of Linaqua and mix until uniform (5 min).

3. Add 725 lb of pigment and mix until smooth and uniform (15-30 min).

4. Reduce with 150 lb of Linaqua and 150 lb of water to grinding consistency.

5. Pass the paste through a Morehouse Mill set for a 3-4 grind.

6. To the letdown tank, charge the remaining 50 lb of Linaqua and 9.5 lb of driers. Thoroughly mix the driers into the Linaqua before the mill base or additional water is added.

7. Mix the grinding paste with the remainder of the vehicle plus driers. Adjust the viscosity with the remaining 75 lb of water, filter, and package.

References

1. SCHLOZ, Can adian Patent 317 827 (1931).
2. H. HADERT, *Casein and Its Uses*, Chemical Publishing, New York, 1938.
3. *Ibid.*
4. H. BENNETT, *Chemical Formulary*, vol. 5, p. 341, Chemical Publishing, New York, 1941.
5. *Ibid.*
6. Personal Communication.
7. BENNETT, *op. cit.* vol. 4, p. 327, 1939.
8. Personal Communication.
9. BENNETT, *op. cit.* vol. 1, p. 269, 1933.
10. *Ibid.*
11. Personal Communication.
12. BENNETT, *op. cit.* vol. 1, p. 270, 1933.
13. *Ibid.* vol. 4, p. 300, 1939.
14. *Ibid.* p. 297, 1939.
15. Personal Communication.
16. BENNETT, *op. cit.* vol. 3, p. 52, 1936.
17. N. J. Zinc Co., New York, N.Y.
18. N. J. Zinc Co., New York, N.Y.
19. N. J. Zinc Co., New York, N.Y.
20. N. J. Zinc Co., New York, N.Y.
21. BENNETT, *op. cit.* vol. 11, p. 221.
22. *Ibid.* p. 226.
23. *Ibid.* p. 229-230.
24. *Ibid.* p. 230-233.
25. *Ibid.* vol. 12, p. 472-473.
26. Baker Castor Oil Co., courtesy of Dewey Almey.
27. Baker Castor Oil Co., courtesy of Union Carbide.
28. BENNETT, *op. cit.* vol. 12, p. 473.

29. *Ibid.* p. 460.
30. *Ibid.* vol. 11, p. 233.

chapter 13

POLISHES

All polishes have several formulation considerations in common. The basic function of a polish is the deposition of a thin, protective or glossy layer on the surface of an article. Since this layer is commonly expected to impart a shine in addition to being water repellant, waxes are a common ingredient. More protection and a deeper shine are created through the use of high-molecular-weight waxes or synthetic resins.

To aid in obtaining a uniform, easy-to-spread, thin film of these high-molecular-weight materials, silicones and fluorocarbons have been used. Fluorocarbons are used as leveling agents in floor polishes. The silicones, because of their incompatability and water repellancy, have been widely used in furniture and automobile polishes. These incompatible materials act as lubricants to allow easier application and also add to the shine and water-repellancy of the polish film. The use of the silicones has become so universal that some polishes presently marketed are little more than silicone emulsions.

Technically, a polish should possess the following characteristics:

1. Ease of application;
2. Ease of "rub-out" for buffable products;
3. Depth of shine;
4. Trueness of color of the polished article;
5. Continuity of shine (lack of ridges, uneven surface);
6. Lack of streaking or smearing when hand or foot is wiped across the surface;
7. Lack of alcohol or water spotting;
8. Durability of the polish coating.

Automobile Cleaner–Polishes

Formula No. 1[1]

A.	Mineral oil (60-80 sec)	6
	Petroleum Naphtha (95-140°C)	6
	Oleic Acid	2
	Carnauba Wax	1
B.	Diatomaceous Earth	11
C.	Amino-Methyl-Propanol	1
D.	Water	73

Melt A and stir in B to make a thick paste. Mix C and D and warm to approximately 75°C. Add CD to the warm melt AB with very vigorous agitation.

The use of a colloid mill for this operation gives excellent results. The proportions of diatomaceous earth, mineral oil, and naphtha may be varied within a reasonable range depending on the grades of the ingredients used and upon the characteristics desired in the cleaner–polish.

No. 2[2]

A.	Silicone Oil*	4.0
	Oleic Acid	2.0
	Mineral Spirits	20.0
B.	Water	47.5
	Triethanolamine	0.2
	Carbopol 934 or 941	0.2
	Snow Floss†	8.0
	Super Floss†	4.0

* DC-200 (350 cS) *Dow Corning Corp.*, Midland, Michigan.

† *Johns-Manville, Celite Div.*, 270 Madison Avenue, New York 10016, New York.

Carefully disperse the Carbopol resin in the water, neutralize with the triethanolamine, and add the abrasives. Dissolve the silicone oil and oleic acid in the mineral spirits. Add A to B with good mixing, using a high-shear agitator such as an Eppenbach mixer.

No. 3[3]

Foam Cleaner-Polish

A.	Dow Corning 200 Fluid, 350 cS	6.0
	Carnauba Wax, No. 1 yellow	2.0
	Stoddard Solvent	18.0
	Kerosene	2.0
	Stearic Acid (triple pressed)	2.5
B.	Triethanolamine	2.5
	Water	16.0
C.	Snow Floss	14.0
	Water	36.8
D.	CMC-7H	0.2

Heat A to about 175°F. to melt the wax. (**WARNING:** Flammable). Mix B and C separately. While mixing A with a mixer with high-shear action (Eppenbach mixer), add phase B slowly.

Maintain the temperature between 160 and 175°F. Remove heat, then add part C and D in turn, with mixing.

No. 4[4]

A.	Dow Corning 200 Fluid, 100 cS	1.5
	Dow Corning 200 Fluid, 1000 cS	4.5

	Carnauba Wax No. 1 yellow	1.0
	Beeswax	1.0
	Oleic Acid	2.5
	Kerosene	2.0
	Stoddard Solvent	18.0
B.	Morpholine	1.5
	Water	16.0
C.	Snow Floss	7.0
	Celite 315	7.0
	Carbopol 934	0.1
	Water	37.9

Heat ingredients of A to 175°F to melt the waxes (**WARNING:** Flammable). While mixing A with a mixer with high-shear action (Eppenbach mixer), add phase B slowly, maintaining the temperature at 160-175°F. Remove heat, then add phase C with mixing.

No. 5[5]

A.	Dow Corning 200 Fluid, 350 cS	4.0
	Stoddard Solvent	19.0
	Kerosene	2.0
	Oleic Acid	2.5
B.	Morpholine	1.5
	Water	16.0
C.	Snow Floss	14.0
	Water	41.0

While mixing A with a mixer with high-shear action, (Eppenbach mixer), slowly add B. Continue the mixing while adding C to the AB combination.

If desired, the oleic acid may be mixed with the ingredients of B. A is then added to B with high-speed mixing. Continue the mixing while adding C.

Automobile Polish*

No. 6[6]

Carnauba Wax	9.0
Beeswax	8.0
Naphtha	75.0
Triethanolamine	2.7
Stearic Acid	7.0
Water	75.0

* About 25 lb of water-absorbing abrasive such as bentonite can be added to produce a paste polish; 60 lb. of an oil-absorbing abrasive such as tripoli makes a liquid polish.

No. 7[6]

Carnauba Wax	9.0
Beeswax	8.0
Naphtha	75.0
Monoethanolamine	1.2
Stearic Acid	7.0
Water	75.0

No. 8[6]

Carnauba Wax	9.0
Beeswax	8.0
Naphtha	75.0
Morpholine	1.7
Stearic Acid	7.0
Water	75.0

"A steam- or hot-water-jacketed kettle is preferred for making wax polishes, as a satisfactory temperature must be maintained to prevent caking of the wax along the sides of the kettle and to avoid discoloration by overheating the wax. A paddle-type, hand-operated stirrer or a low-speed, large-bladed propeller is also suggested for successful operation. Since morpholine has a flash point of 100°F, it should not be added to the mixture in the presence of open flames. If the wax is melted by means of a gas burner, the gas should be turned off during the addition of the morpholine.

"Melt the waxes and stearic acid, add the amine, and maintain the temperature at about 90°C. Add the naphtha slowly and stir until a clear solution is obtained and the temperature is 90—95°C. Avoid the use of open flames.

"The method of adding the abrasive depends on the type used. An oil-absorbing abrasive such as tripoli should be well mixed with the hot naphtha solution of waxes just before the water is added. An abrasive that absorbs water, such as bentonite, is best stirred into the finished emulsion.

"Heat the water to boiling, add it to the naphtha solution, and stir vigorously until a good emulsion is obtained. Continue stirring slowly until the emulsion has cooled to room temperature.

"The proportions of waxes can be changed as desired, depending on the ease of polishing required and the hardness of the final film. A high-melting hydrocarbon wax can be used in place of all or part of the beeswax with good results. When the primary use of the automobile polish is for polishing rather than as a cleaning and polishing combination, it will be more satisfactory without an abrasive."[6]

Auto Wax-Wash Polishes

No. 9[7]

Phase I	
A. Flexowax C	10.4
Oleic Acid	1.7
B. Triethanolamine	0.7
Water	7.0
C. Water	49.8
Phase II	
D. Dow Corning 200 Fluid (350 cS)	8.75
Acintol FA-2	0.50
Triethanolamine	0.25
E. Ammonia (26°Bé)	0.15
Water	15.35
CMC 7H	0.40
Triton X-100	5.00

While mixing A at 195°F in a high-speed, high-shear mixer (Eppenbach), slowly add B. Maintain the temperature during the addition. Finally, add C to AB with continued agitation. Cool and filter.

To prepare Phase II: While mixing D in a high-speed, high-shear agitator, slowly add E.

While blending Phase I with Phase II with a propeller-type agitator, slowly sift the dry CMC into the vortex so that it does not lump. Continue agitation until the CMC is dissolved. Mix in the Triton X-100 and package.

No. 10[8]

Silicone Fluids	20.0
Surfactol 318	9.8
Paraffin Oil	7.5
Triethanolamine	1.2
Ammonia (26° Bé)	0.7
Water	60.8

"Dry-Bright" Floor Polishes

No. 11[9]

Carnauba Wax No. 3, refined	6.0
Microcrystalline Wax	3.0
Emulsifiable Resin	3.0
Oleic Acid	1.44
Amino-Methyl-Propanol	0.91
Water	85.65

Leveling agent, 12% solids 20% or more of above

Melt the wax and resin together (about 320-350°F, depending on the resin), cool to about 220°F, and add the oleic acid. Continue cooling to about 200°F, add the amino-methyl-propanol with good stirring and continue stirring for about 5 minutes, keeping the temperature at about 200°F. Heat the water to about 200-210°F and add the wax–resin mixture to it, agitating thoroughly during the addition. When the product has cooled to room temperature, a leveling agent may be added. It is then ready for packaging.

No. 12[10]

A.	Rhoplex B-231 (40%)	80
	Durez 19551 (30.8%)	13
	PolyEm 20 (40%) (*Spencer Co.*)	10
B.	Carbitol (*Union Carbide*)	4
	FC 128 (1%) (3*M Corp.*)	1
	Ethylene Glycol	4
	Tributoxyethyl Phosphate	1
	Triton N-100 (*Rohm and Haas*)	0.8

Blend A with good agitation and add B slowly to A. Stir AB for at least ½ hour, then add 500 ppm formaldehyde as preservative.

No. 13[11]

Wax Emulsion

Carnauba	30
A-C Polyethylene 629	10
Oleic Acid	7
Morpholine	6
Water	260
Leveling agent solution	
Shanco L-1001	30
Borax, 5 mol	13
Water	244

Heat the water to 194°F and add the borax with agitation until dissolved. Maintaining the temperature at 194°F, add the Shanco resin and stir until the solution is achieved.

No. 14[12]

A.	Duroxon J-324	150
	Oleic Acid	10
	Morpholine	17
	Monoethanolamine	3
	Water to make	100

(Appx. 16% solids)

B.	Shanco L-1001	160
	Ammonia (28%)	36
	Water	804

Final composition

A. 85 parts (by volume)
B. 15 parts (by volume)

This product can be readily made by conventional procedures. When the water-to-wax method of manufacture is used, it can be modified by reducing the amount of amine (morpholine and mono-ethanolamine) recommended for A. It is also possible to to employ leveling resins other than the re-commended Shanco L-1001.

No. 15[12]

A.	Duroxon J-324	100.0
	Oleic Acid	14.0
	Morpholine	14.3
	Monoethanolamine	3.1
	Water to make (16% solids)	712.0
B.	Ubatol U-2003 at 40%	100.0
	Plasticizer KP-140	2.6
	Dibutyl Phthalate	3.3
	Water	46.6
C.	Durex 15546 resin	140.0
	Ammonia (28%)	21.8
	Water	838.2

Final Composition (add in the order listed):

A	36.5
Water	26.0
B	23.5
C	14.0

It is recommended that A be prepared by the wax-to-water method. This emulsion should be almost completely transparent. Best leveling is usually obtained after the final composition has been allowed to stand undisturbed for at least 24 hours.

Other Floor Polishes

No. 16[13]

Chlorowax 70	17.0
Crown Wax 23	66.0
Oleic Acid	8.3
Morpholine	9.0
Water	730.0

Melt the waxes together and add the oleic acid and morpholine with stirring. Maintain the gel at 200—210°F and add hot water at 205—210°F, slowly at first, with rapid stirring. After the gel inverts to an oil-in-water emulsion, the water be added more rapidly. When half the water has been added, discontinue the heat and cool the batch as rapidly as possible while the remainder of the water is added at room temperature, with slow agitation.

No. 17[13]

Chlorowax 70	7.7
Carnauba Wax	5.8
Crown Wax 23	64.0
Oleic Acid	7.7
Triethanolamine	9.7
Borax	5.2
Water	730.0

Use the procedure given under Formula 16, but slowly add the borax dissolved in 25 lb of boiling water to the mixture of waxes and emulsifier.

No. 18[14]

Estawax 20 4-5
Paraffin Wax (140°F AMP) 7-6
Stoddard Solvent 89

No. 19[14]

Estawax 25 4-5
Paraffin Wax (140°F AMP) 7-6
Stoddard Solvent 89

Heat the solvent to 180°F. and add the molten waxes. Cool with stirring and pour into containers at 100—140°F. Do not stir rapidly when the mixture approaches the pouring temperature. Rapid agitation will affect the crystal formation adversely and may cause separation of the solvent.

No. 20[15]

A. Duroxon J-324 39.0
Shanco 300 Resin 39.0
Prime Yellow Carnauba Wax 29.0
Oleic Acid 11.0
Morpholine 7.5
Borax 4.5
Potassium Hydroxide 0.4
Water to make 12% solids

Add the melted wax to water.

B. Durez 15546, 12% Ammonia cut

Mix 80 parts of A with 20 parts of B.

No. 21[15]

A. Duroxon J-324 39.0
Shanco 300 Resin 39.0
Duroxon H-110 29.0
Oleic Acid 5.0
Morpholine 11.0
Borax 4.5
Potassium Hydroxide 0.4
Water to make 12% solids

B. Durez 15546, 12% Ammonia cut

Mix 80 parts of A with 20 parts of B.

No. 22[15]

A. Duroxon J-324 20.0
Oleic Acid 2.0
3-methoxy-propylamine 1.5
Water to make 13% solids

B. Ludox, reduced to 13% solids

C. Manila Loba C, resin dispersion 13%

Mix 50 parts of A with 25 parts of each B and C.

No. 23[15]

Duroxon J-324 20
Shanco 300 Resin 20
Oleic Acid 4
Morpholine 6
Water to make 15% solids

No. 24[15]

Duroxon H-110 20
Durez 219 Resin 20
2-amino-2-methyl-1-propanol 6
Water to make 13% solids

No. 25[15]

A. Duroxon H-110 12.0
Oleic Acid 2.4
Morpholine 1.6
(Water to make 15% solids)

B. Ubatol 2001 50

Water	50

Add the melted wax to the water. Mix 60 parts of A with 40 parts of B.

No. 26[15]

Duroxon H-110	50
Morpholine	6
Water	to make 12% solids

Furniture Polishes

No. 27[16]

A.	100/100 Paraffin Oil	37.0
	Monateric S-MEA	12.5
	Oleic Acid	1.0
	Monawet MO-70	0.5
Water		49.0

Blend A together, then add water slowly in small amounts until a definite thickening has occurred. The mass is left standing in this state for a minimum of ½ hour, after which the balance of the water can be added rapidly with vigorous stirring.

No. 28[17]

A.	Silicone oil*	4.0
	Carnauba Wax	1.0
	Cardis Wax†	1.5
	Paraffin Wax	2.5
	Mineral Spirits	20.0
	Span 60††	1.5
	Tween 60††	1.5
B.	Carbopol 934	0.2
	Sodium hydroxide (10% solution)	0.8
Water		67.0

Carefully disperse the Carbopol 934 in the water and neutralize with the sodium hydroxide. Heat both A and B to 60°C (140°F) and add A to B with vigorous agitation. A very white, oil-in-water emulsion forms immediately.

* SF-96 (300) *General Electric Co.*, Silicone Products Dept., Waterford, N.Y.

† *Warwick Wax Co. Inc.*, 10th St. and 44th Ave., Long Island City 1, New York.

†† *Atlas Chemical Industries, Inc.*, Wilmington 99, Delaware.

Aerosol Furniture Polishes

No. 29[18]

A.	Dow Corning 200 Fluid, 350 cS	15.0
	Flexowax C	5.0
	Oleic acid	1.5
B.	Morpholine	1.0
	Water	17.5
C.	Water	60.0

Heat A and B to 175°F. While mixing in a high-shear mixer (Eppenbach), slowly add B to A. With continued mixing, add C and cool to room temperature. Pressure fill.

Suggested aerosol formulation

Concentrate (above)	18.0
Water	72.0
Propellent 114	5.0
Propellent 12	5.0

Can: Lacquer-lined

Valve: No. 1 Precision, 0.018" body, 0.018" stem

Button: 0.016" MBRT

No. 30[18]

Concord Co-Wax, 10% emulsion	2.2
Dow Corning EF-1-0176A	7.2

Water	90.6

Thoroughly mix and pressure fill using the indicated propellants as below.

Suggested aerosol formulation

Concentrate (above)	90.0
Propellent 114	8.0
Propellent 12	2.0

Can: Lacquer-lined, 6 oz
Valve: No. 1 Precision, 0.018" body, 0.018" stem
Button: 0.016" MBRT

Paste Polishes

No. 31[19]

Carnauba Wax	30.0
Beeswax	30.0
Naphtha	50.0
Triethanolamine	4.3
Stearic Acid	8.0
Water	65.0

No. 32[19]

Carnauba Wax	30.0
Beeswax	30.0
Naphtha	50.0
Monoethanolamine	1.9
Stearic Acid	8.0
Water	65.0

No. 33[19]

Carnauba Wax	30.0
Beeswax	30.0
Naphtha	50.0
Morpholine	2.6
Stearic Acid	8.0
Water	65.0

No. 34[20]

A. Carnauba Wax	14.0
Beeswax	14.0
Paraffin Wax	14.0
Stearic Acid	5.0
B. Amino-methyl-propanol	2.5
Water	80.0

Melt A together on a boiling water bath. Stir B into the wax melt (temperature should not exceed 100°C to 105°C). Then add the boiling water to AB with vigorous agitation.

No. 35[21]

Carnauba Wax	12.0
Beeswax	6.0
Naphtha	70.0
Triethanolamine	4.8
Stearic Acid	8.0
Water	180.0

No. 36[21]

Carnauba Wax	12.0
Beeswax	6.0
Naphtha	70.0
Monoethanolamine	2.1
Stearic Acid	8.0
Water	180.0

No. 37[21]

Carnauba Wax	14.0
Beeswax	4.0
Naphtha	25.0
Monoethanolamine	2.0
Stearic Acid	8.0
Water	240.0

No. 38[21]

Carnauba Wax	12.0
Beeswax	6.0
Naphtha	70.0
Morpholine	3.0
Stearic Acid	8.0
Water	180.0

Mineral Oil Emulsion Polishes

No. 39[22]

Mineral Oil (light)	40.0
Ethofat 60/15	2.5
Ethofat 60/20	2.5
Water	55.0

Dissolve the Ethofat 60/15 and Ethofat 60/20 in the mineral oil using heat if necessary. The water is then added to the oil with agitation.

No. 40[23]

Light Mineral Oil	48.0
Surfactol 365	8.0
Oleic Acid	6.6
Monoethanolamine	0.5
Water	60.0

No. 41[24]

Silicone Polishing Cloth*

A. Methypolysiloxane Oil	12
Isopropanol	6
Triethylamine	7
Oleic Acid	1
B. Cresol Soap Solution (1%)	175

Mix A and pour into B. Impregnate a soft cotton cloth with the above for 15 minutes; squeeze and dry.

* German Patent 941,309

No. 42[25]

Aerosol Glass Cleaner-Polish

A. Triton W-30	0.2
Isopropanol (99%)	2.0
Dowanol EE	0.3
Water	96.0
B. Dow Corning EF-1-0108	1.5

Mix A (**WARNING: Flammable**). When homogeneous, add A to B with gentle mixing. Package as indicated below.

Suggested aerosol formulation

CP-7-0048A Concentrate	90
Propellent 114	6
Propellent 12	4

Can: Lacquer-lined
Valve: Risdon 5210, EH-6 spray tip

NOTE: Do Not Shake. Hold container upright about six inches from window to be polished and spray lightly. Wipe dry with a soft clean cloth.

No. 43[26]

Metal Polish

Condensate PHL	4.0
Silica	17.8
Water	78.0
Cottonseed Fatty Acid	0.2

No. 44[27]

Silver Polish

(Lotion type)

A. P and G Amide No. 72	5.00
Red Oil (Oleic Acid)	3.25
B. Water	70.75
C. P and G Amide No. 23	2.00
D. Trisodium Phosphate (Dodechydrate)	2.00
Thiourea	2.00
Super Floss	15.00

Blend components of A together and after these are thoroughly mixed, add the water (B) and stir until completely dissolved. Then add C, heating to about

140°F and stirring until it is dissolved. Dissolve D in ABC and finally add the Super Floss, stirring until a smooth creamy lotion is obtained.

No. 45[27]
(Paste type)

A.	Amber Granules	5
B.	Soda Ash (Sodium Carbonate)	3
	Diatomaceous Earth*	10
	Water	82

Dissolve A in half the water at approximately 120°F. Separately, add B to the remaining water and mix to dissolve the soda ash. It is desirable to heat the B mixture in order to drive off any entrained air. Thoroughly mix A and B.

* A soft, fine abrasive variously known as kieselguhr, guhr, infurorial earth, fossil flour, tripolite, ceyssatite, and diatomite. Also sold under proprietary names such as Celite.

References

1. Commercial Solvents Corp.
2. B. F. Goodrich Chemical Co.
3. Wyandotte Chemicals Corp.
4. Wyandotte Chemicals Corp.
5. Wyandotte Chemicals Corp.
6. H. BENNETT, *Chemical Formulary*, vol 11, p. 267, Chemical Publishing, New York.
7. Wyandotte Chemicals Co.
8. Baker Castor Oil Co.
9. Commercial Solvents Corp.
10. Rohm and Haas Co., *Soap and Chem. Specialties*, 41:203 (1965).
11. BENNETT, *op. cit.* vol. 11, p. 263-264.
12. *Ibid.* p. 272-273.
13. *Ibid.* p. 279.
14. *Ibid.* p. 267-268.
15. *Ibid.* p. 268-269.
16. Mona Industries, Inc.
17. B. F. Goodrich Chemical Co.
18. Wyandotte Chemicals Corp.
19. BENNETT, *op. cit.* vol. 11, p. 265-266.
20. Commercial Solvents Corp.
21. BENNETT, *op. cit.* vol 11, p. 266.
22. *Ibid.* p. 273.
23. Baker Castor Oil Corp.
24. BENNETT, *op. cit.* vol. 11, p. 273.
25. Wyandotte Chemicals Corp.
26. Continental Chemical Co.
27. Proctor and Gamble, Co.

References

chapter 14

RESIN AND RUBBER EMULSIONS

Rosin Emulsions

Formula No. 1[1]

Rosin	70.0
Water	210.0
Glue	15.0

Melt the glue in water and while boiling, slowly add the melted rosin, agitating violently. Continue agitation until perfectly smooth.

No. 2[2]

Rosin	70.0
Water	210.0
Gelatin	15.0

Melt the gelatin in water and while boiling add the melted rosin slowly, agitating violently. Continue agitation until perfectly smooth.

No. 3[3]

Rosin	70.0
Water	210.0
Stearic Acid	6.3
Triethanolamine	2.1

Melt the rosin and stearic acid together. Add the triethanolamine to the water. Heat the water to boiling point and stir in the melted rosin until smooth.

No. 4[4]

A.	Rosin	10.00
B.	Naphtha	10.00
C.	Ammonium Linoleate	0.30
D.	Ammonium Hydroxide	0.25
E.	Water	20.00

Heat A to 150°C and turn off the flame; run B, which has been previously heated on a water-bath to 90°-100°C, into it slowly and stir until all the rosin has dissolved. Cool and add slowly with vigorous stirring C, D, and E mixed together.

No. 5[5]

Rosin, Ester Gum, Cumar, or other resin (melted)	3 lb
Oleic Acid	1 oz
Triethanolamine	2 oz
Casein Solution	2 lb
Water	to make 1 gal

No. 6[6]

Rosin–Turpentine Emulsion

Rosin	11.0
Turpentine	2.5
Ammonium Linoleate	2.0
Water	50.0
Ammonia	15.0

The ammonium linoleate and water are taken up in the usual way, heated, and mechanically agitated (high-speed mixer). The rosin and turpentine are heated together and added to the ammonium linoleate dispersion in water to which the ammonia has previously been added. Stirring is continued until cool. This gives a paste emulsion.

No. 7[7]

Paraffin–Rosin Emulsion

In producing these dispersions, various proportions of paraffin and rosin may be used—from 85% paraffin and 15% rosin to 15% paraffin and 85% rosin. Assuming that a mixture of about 50% paraffin and about 50% rosin is employed, the mixture is melted and heated to about 220°F, which temperature is materially above the melting points of paraffin (130°F) and rosin (about 180°F.). The heat-liquefied thermoplastic mixture is then comingled as a regulated stream flowing at 15 1b/min with a regulated stream of 5% caustic soda solution at 110°F flowing at 20 1b/min.

Under these conditions and at a temperature of about 155°F, an aqueous dispersion of a solids content of about 45% is produced that is of a creamy consistency when suddenly chilled to a temperature below 130°F to prevent coalescence of dispersed particles.

No. 8[8]

Asphalt–Rosin Emulsion

In producing asphalt dispersions, at least about 35% rosin, based on the total weight of thermoplastic material, should be used in order to produce a stable dispersion of fine particle size. Assuming that a mixture of about 50% asphalt (with a melting point of about 150°F) and about 50% rosin is employed, the mixture should be heated to about 300°F, whereupon it is brought in contact with a caustic soda solution under practically the same conditions as in the case of the paraffin dispersion, except that the solution should be at 150—160°F. The resulting dispersion has a temperature of about 200°F and when permitted to cool is of a smooth, soft, paste-like consistency.

No. 9[9]

Cumar Emulsion

A.	Naphtha	20
	Cumar	20
B.	Ammonium Linoleate Paste	10

Water 40

Heat A until dissolved. While at 90-100°C, add B (at 90-100°C) slowly with vigorous stirring.

No. 10[10]

Gelva Resin Emulsion

Polyvinylacetate Resin (Gelva)	25.0
Toluol	55.0
Butyl Acetate	20.0
Water	39.4

Sodium Lauryl Sulfate	0.2
Sulfonated Castor Oil	0.4

No. 11[11]

Pentaerythritol Abietate Emulsion

Pentaerythritol Abietate	50.0
Oleic Acid	20.0
Triethanolamine	4.2
Mineral Spirits	15.0
Water	500.0

Staybelite Ester Emulsions

No. 12, 13, 14, 15[12]

	Method A		*Method B*	
	12[12]	13[12]	14[12]	15[12]
Staybelite Ester No. 10, 1, or 2 (80% solution*)	400.0	---	47.5	---
Staybelite Ester No. 3 (solids basis)	---	80.0	---	42.0
Duponol ME	0.8	2.0	---	---
Potassium Hydroxide	---	---	0.3	0.3
Oleic Acid	---	---	1·1	1.2
Sulfated Castor Oil (75%)	0.8	2.0	0.9	1.0
Water	138.4	136.0	50.2	55.5
% Solids	37	38	40	45

Emulsions of Staybelite esters may be prepared by two general methods, A and B. Method A requires the use of a colloid mill and only a minimum of emulsifier. Method B yields spontaneous emulsions on rapid agitation, but requires the use of large amounts of a soap such as potassium oleate. Staybelite Esters No. 10, 1, and 2 are introduced as 80% solvent solutions; Staybelite Ester No. 3 is so fluid that it may be emulsified without the aid of solvent.

Method A. Blend the solvent solution of the Staybelite Ester (in the case of Staybelite Ester No. 3, where no solvent is used, warm to 90°C before blending) with the water containing the Duponol ME and sulfated castor oil, then pass through the colloid mill.

Method B. Stir the oleic acid into the solvent solution of the Staybelite ester (or into the resin itself when Staybelite Ester No. 3 is used).

* In Varsol, HF naphtha, toluene, xylene, tollac, Solvesso No. 2, Solvent 50D, etc.

Dissolve all the potassium hydroxide and sulfated castor oil in half the water. Add the resin–oleic acid mixture to this solution with vigorous agitation to produce a spontaneous emulsion. Stir vigorously for 10-15 minutes, dilute with the remaining water, stir 30-40 minutes, filter. The emulsion is then ready for use.

Ethyl Cellulose Emulsions

No. 16, 17, 18

		16[13]	17[14]	18[15]
A.	Ethyl Cellulose	5.0	5.0	5.0
	Octyl Acetate	63.0	5.0	5.0
	Toluol	63.0	63.0	63.0
	Diphenyl Oxide	0.6	1.4	1.0
	Dow Plasticizer No. 6	1.4	0.6	1.0
B.	Deceresol O. T.	2.0	2.0	2.0
	Water	13.0	13.0	13.0

Add B to A slowly while mixing vigorously with an electrical mixer.

No. 19[16]

Laminating Adhesive

25% Solids high-viscosity Ethyl Cellulose dissolved in 80:20 mixture of Toluene and Ethyl Alcohol	100
Stabelite Ester No. 3	20
Water containing 1 part 28% Ammonia	30

The Staybelite Ester No. 3 is dissolved in the ethyl cellulose solution. To this resin-modified solution the water phase is added with slow agitation to produce a homogeneous stable water-in-oil reversed emulsion.

Methyl Methacrylate Emulsions

No. 20[17]

A.	Monomeric Methacrylate	25
B.	Dibutyl Phthalate	5
C.	Deceresol O. T. (10%)	4
D.	Water	64
E.	Hydrogen Peroxide (30%)	3

Mix A and B and add slowly to the mixture of C, D, and E, stirring well. Pass through a colloid mill or homogenizer.

No. 21[18]

A.	Monomeric Methacrylate	12.5
	Monomeric Styrene	12.5
B.	Dibutyl Phthalate	5.0
C.	Deceresol O. T. (10%)	4.0
D.	Water	64.0
E.	Hydrogen Peroxide	2.0

No. 22[19]

Methyl Methacrylate	30.0
Water	100.0
Ammonium Persulfate	0.3

Heat to 80°C, and mix vigorously.

No. 23[20]

Flexible coating for cloth)

Methyl Methacrylate	2.50 lb
Potassium Persulfate	0.05 oz
Soap	5.60 oz
Water	10.00 lb

The materials are placed in a closed container and agitated at 50°C for 1 day or may be stirred in an open vessel fitted with a reflux condenser. The product is an emulsion of a flexible polymer. Cloth is coated by passing it through this emulsion. The dried coating is resistant to water and many organic solvents.

Natural Resin Emulsions

No. 24[21]

Elemi	58
Batavia Dammar	14
Water	128
Triethanolamine	3
Oleic Acid	7
Commercial Latex Solution (about 36% latex)	70

Melt the Elemi and stir in the Dammar. Raise the temperature to 125°C. When the mixture is homogeneous, allow it to cool and set the container into a boiling water bath. Add the oleic acid and then the triethanolamine with stirring. Add the water, preheated to boiling, portion by portion, stirring and incorporating after each small portion has been added before adding the next portion. The mixture thickens at first, but as additions of water continue, it changes to a thin white liquid. After all the water has been added, allow the mixture to cool, but continue the stirring until it is cool. Add the latex solution with stirring.

No. 25[22]

Manila DBB (ground)	14.4
Water	20.6
Ammonium Hydroxide	6.0
Commercial Latex Solution (about 36% latex)	369.0

Mix the Manila, water, and ammonia until all the resin dissolves; strain. To this solution, add the latex solution.

No. 26[23]

Manila DBB	14.4
Water	20.6
Ammonium Hydroxide	6.0
Commercial Latex Solution (about 36% latex)	123.0

Procedure as in Formula No. 25.

No. 27[24]

Cellulose Ester Emulsion

A pyroxylin base is prepared by colloiding 12.5 parts by weight of alcohol-wetted pyroxylin (10 parts of dry $\frac{1}{2}$ sec pyroxylin) with 20 parts by weight of blown linseed oil in a suitable mixer, such as the Werner–Pfleiderer mixer. 25 parts by weight of a solvent mixture are then added to the colloided mass in portions of 5 parts by weight to form a homogeneous base with the following composition:

Pyroxylin ($\frac{1}{2}$ sec)	10.0
Alcohol (denatured)	2.5
Blown Linseed Oil	20.0
Butyl Acetate	20.0
Butyl Lactate	5.0

An emulsion is prepared by heating 0.5 part by weight of sodium oleate with 15 parts by weight of gasoline to a clear gel, after which 2 parts by weight of water are added to the hot gel with vigorous stirring, thus forming a concentrated emulsion of gasoline in water, stabilized by sodium oleate. For convenience this will hereafter be called the agent emulsion.

The presolution or solvation of the sodium oleate in gasoline or similar liquid is desirable to assure uniform distribution.

Into 57.5 parts by weight of the pyroxylin base, 17.5 parts by weight of the agent emulsion are then stirred vigorously with a high-speed stirrer of the propeller-blade type.

No. 28[25]

Polystyrene Resin Emulsion

A.	Polystyrene Resin	7.5
	Toluol	22.5
	Octyl Acetate	6.5
B.	Emulsifying Agent (Deceresol O. T.)	1.0
	Diglycol Laurate	0.5
	Water	12.5
C.	Toluol	50.0

Dissolve A. To this add B slowly while mixing vigorously; then add C and stir until uniform. This gives a stable film-forming emulsion.

Vinylite Emulsions

"One of the most satisfactory emulsifying agents is ammonium oleate, prepared *in situ* by adding oleic acid to the lacquer phase and an excess of ammonia to the water phase. Triisopropanolamine oleate prepared in a similar manner can be used if the volatility of ammonia is undesirable. Sodium lauryl sulfonate may also be used and this is dissolved entirely in the water phase. The amount of emulsifying agent required is of the order of 0.5-5% of the weight of resin present, the exact amount varying for specific agents and specific types of emulsion.

"A small amount (0.5% of resin weight) of bentonite clay dispersed first in a little water and then added to the finished emulsion improves appreciably the stability upon storage. It should not be added to the water phase before emulsification. A ball mill is suggested as a suitable means of preparing the preliminary bentonite dispersion."[26]

No. 29[26]

(Plasticized Vinylite Resin AYAA Emulsion)

Vinyl Resin AYAA	50.0
Tricresyl Phosphate	5.0
Toluene	43.5

Oleic Acid	1.5
Distilled Water	92.0
Aqua Ammonia	8.0

No. 30[27]

(Emulsion Polymerization of Vinyl Acetate)

A.	Water	50.00
	Aerosol OT	2.50
	Potassium Metabisulfite	0.05
B.	Vinyl Acetate	10.00

Dissolve A, add B, and stir. Heat for 1 hr at 60-65°C.

Rubber Emulsions

No. 31[28]

Rubber	70.00
Bentonite	1.25
Water	10.75

No. 32[29]

A mixture of 100 parts rubber, 10-100 parts benzene, 1 part glue, and 1 part casein is masticated, while an aqueous solution of potassium oleate is slowly added until the rubber constitutes the dispersed phase of the batch. The product is suitable for use as a cementing medium.

No. 33[30]

(Artificial Latex)

Reclaimed Rubber	48.5
Cumar, mineral rubber of gum rosin	2.5
Rosin Soap	9.0
Whiting	40.0
Water	to suit

The reclaimed rubber and softener are mixed together in a Banbury mixer for approximately 20 minutes. The whiting and rosin soap are made into a stiff putty-like paste with water. This paste is then placed in a Werner–Pfleiderer mixer and the rubber–softener blend added gradually to the paste in the mixer. This process will require 30 to 80 minutes, during which time the mixer should be heated with steam passing through the jacket to a temperature of approximately 180°F. The mixing is continued until the rubber is completely dispersed in the form of fine particles 2-20μ in diameter. The dispersion thus made can be diluted with water as desired to reduce the viscosity. Bodying agents (such as gum tragacanth) and curing ingredients may be added with the water.

No. 34[31]

(Synthetic Rubber Emulsion Polymerization)

Butadiene	7.50
Styrene	2.50
Isohexyl Mercaptan	0.05
Water	18.00
Sodium Oleate	2.00
Ammonium Persulfate	0.03

Polymerize at 30°C for several days until the desired degree of polymerization is achieved.

No. 35[32]

Rubber Resin Emulsion

A.	Rubber Resin	40
B.	Ammonium Linoleate Paste	10
C.	Water	150

Heat A and B until homogeneous and add C slowly with vigorous mixing.

No. 36[33]
(Insole Cement)

Latex (40% Solids)	100
Casein Solution (10%)*	10-15
Resin Emulsion (50%)†	10

**Casein Solution*

Casein	5 lb
Water	44 lb
Ammonium Hydroxide	8 oz
Phenol	1 oz

†*Resin Emulsion*

Resin (Ester Gum or Cumar)	3 lb
Oleic Acid	3 oz
Ammonium Hydroxide	1 oz
Water	3 lb

The oleic acid and ammonium hydroxide are dissolved in the water and the resin is emulsified by heating to the melting point and stirring vigorously.

Chlorinated Rubber Emulsions

No. 37[34]

Stable emulsions can easily be obtained without the use of emulsifying agents—for example, with 70 g of chlorinated rubber, 30 g of xylol, and 40 g of water, and by feeding each of these simultaneously and progressively into a stirring device.

No. 38[35]

A.	Casein	2.0
	Ammonia (28%)	0.1
	Turkey Red Oil	6.0
B.	Chlorinated Rubber	20.0
	Trichlorethylene	80.0

Heat A to 35°C with stirring and add slowly a solution of B. Stir until uniform. This disperses well in water.

No. 39[36]

20 parts by weight of oleic acid, saponified with 20 parts of sodium silicate in 200 parts of water, are stirred at 100°C into 5 parts of chlorinated rubber dissolved in 25 parts of resin oil; 125 parts of water containing 8 parts of casein and 0.5 parts of ammonia are then added.

No. 40[37]

Chlorinated Rubber	100.0
Hydrogenated Methyl Abietate	15.0
Oleic Acid	4.9
Toluol	16.0
Xylol	110.0
Water	98.5
Caustic Soda	0.7

Emulsion Adhesives

No. 41[38]
(40% Resin Solids)

Hard "Neville" Resin	42
Solvent	18
Oleic Acid	4
Triethanolamine	2
Water	40

Initially, the resin is dissolved in the solvent. The oleic acid and triethanolamine are then added to the resin solution and mixed until uniform using an "Eppenbach Homomixer," "Cowles Dissolver," or some other high-speed mixer. The water is then added in small portions under high-speed mixing. At first, a water-in-oil emulsion is formed with a noticeable increase in viscosity. However, with the continued addition of water, the emulsion will reach a maximum viscosity and invert to an oil-in-water type with some decrease in viscosity. After inversion, the remainder of the water can be added more rapidly. This emulsion is water dispersible and can be further reduced with water to any desired solids concentration.

No. 42[39]
(48% Resin Solids)

"*Nevillac*" Hard	60.0
1:1 vol. mixture of Mineral Spirits-Toluol	15.0
Aerosol OT (75%)	2.4
Tergitol NPX	2.0
Water	44.0

Initially, the resin is dissolved in the solvent mixture. The surfactants are then added to the resin solution and mixed until uniform. The water is added with continued high speed mixing. The resultant oil-in-water emulsion can be thinned as desired.

No. 43[40]
(47% Resin Solids)

Hard Neville Resin	45
Methocel, 400 cP (3% solution)	30
Tamol 731 (25%)	2
Water	20

Charge resin, Tamol 731, and water into a ball mill and mill until a desired particle size is obtained. The Methocel solution can then be added.

No. 44[41]
(37% Resin Solids)

Hard Neville Resin	300.0
Water	500.0
Triton NE	1.5
Tween 20	1.5

The resin is initially powdered by ball milling or any other conventional means. A slurry is made of the above and passed through a colloid mill at 7200 rpm (Premier Mill, 3½ in. pulley). Multiple passes are made with mill setting reduced from 0.010 in. to 0.002 in.

No. 45, 46[42]
Shoe Adhesive
(Vulcanizing Type)

	Wet 45[42]	*Dry* 46[42]
Natural Rubber Latex	167.0	100.0
Potassium Hydroxide (10% solution)	10.0	0.5
Ammonium Caseinate (10% solution)	10.0	0.5
Zinc Oxide	----	2.0

Sulfur	----	1.0
Nevastain B	----	2.0
Neville LX-685, 125, Resin Solution Emulsion	187.0	75.0
Setsit-5	3.0	1.5

The above ingredients are slowly stirred into the natural rubber latex in the order shown. The zinc oxide, sulfur, and Nevastain B are combined and ball-milled at about 50% solids using a suitable dispersing agent, following the general formula and technique shown for Emulsion Type Adhesive, Formula No. 44. The resultant dispersion of these ingredients is slowly stirred into the batch along with the Neville LX-685, 125, Resin Solution Emulsion, prepared according to Emulsion Type Adhesive, Formula No. 41. Finally, the accelerator is added after being diluted with an equal volume of water.

A suggested cure for this adhesive would be 20 minutes at 300°F.

No. 47[43]

Silicone Mold Release

A.	Triton X-100	2.00
	Dow Corning 200 Fluid, 350cs.	35.00
	OPE 30 (Octylphenoxy-ethanol)	1.00
	Triton W-30	0.50
B.	Water	61.50

Mix A and add B slowly with high speed agitation (e.g. in a colloid mill) until a "thick phase" is formed. Mix the thick phase thoroughly until it is uniform. Add the remainder of the water slowly with continued agitation.

Use at 1 part above dilution with 70 parts of water for spray-on mold release. Wetting can be improved by addition of 1-2% Renex 698 to the starting emulsion.

No. 48[44]

Silicone Oil Emulsion

DC-200 Silicone Oil (*Dow Corning Corp.*)	200.0
Carbopol 934	2.0
Dodecylamine	1.0
Sodium Hydroxide (10% solution)	6.0
Water	200.0

Carefully disperse the Carbopol 934 in the water and mix to a thin, cloudy dispersion without lumps. Add the sodium hydroxide and then the amine with mild agitation. The dispersion will gel immediately. Lastly, add the DC-200 in a slow, steady stream with good mixing. A white, stable emulsion forms immediately.

No. 49[45]

Rubber Dressing

Amber Granules	21
Water	57

High Flash Naphtha 19
Lamp Black 3

Wet out the amber in a small amount of cool water and then dissolve in the total amount of water at about 130°F. Make the emulsion by adding the naphtha slowly with constant stirring. After the emulsion forms, stir in the lamp black while the mixture cools.

References

1. H. Bennett, *Chemical Formulary*, vol. 1, p. 155, Chemical Publishing, New York, 1933.
2. *Ibid.*
3. *Ibid.*
4. *Ibid.* vol. 1, p. 443, 1933.
5. *Ibid.* vol. 2, p. 189, 1935.
6. *Ibid.* vol. 1, p. 163, 1933.
7. *Ibid.* vol. 2, p. 189, 1935.
8. *Ibid.*
9. *Ibid.* vol. 2, p. 188, 1935.
10. Billing, US Patent 2 230 792 (1941).
11. US Patent 2 333 887.
12. Bennett, *op. cit.* vol. 7, p. 113, 1945.
13. *Ibid.* vol. 5, p. 73, (1941).
14. *Ibid.*
15. *Ibid.*
16. *Ibid.* vol. 7, p. 40, 1945.
17. *Ibid.* vol. 5, p. 73, 1941.
18. *Ibid.*
19. *Ibid.* vol. 5, p. 550, 1941.
20. *Ibid.* vol. 7, p. 285, 1945.
21. American Gum Importers' Assn.
22. American Gum Importers' Assn.
23. American Gum Importers' Assn.
24. US Patent 1 970 572.
25. Bennett, *op. cit.* vol. 5, p. 72, 1941.
26. Union Carbide Corporation, New York (Catalog).
27. Bennett, *op. cit.* vol. 7, p. 357, 1945.
28. US Patent 1 498 387.
29. Bennett, *op. cit.* vol. 1, p. 163, 1933.
30. *Ibid.* vol. 5, p. 543, 1941.
31. US Patent 2 281 613.
32. Bennett, *op. cit.* vol. 5, p. 72, 1941.
33. *Ibid.* vol. 7, p. 43, 1945.
34. *Ibid.* vol. 4, p. 98, 1939.
35. *Ibid.*
36. *Ibid.* vol, 3, p. 110, 1936.

37. Billing, US Patent 2 230 792 (1941).
38. Bennett, *op. cit.* vol. 11, p. 47.
39. *Ibid.* vol. 11, p. 48.
40. *Ibid.* vol. 11, p. 49.
41. *Ibid.*
42. *Ibid.*
43. Dow Corning Corp.
44. Bennett, *op. cit.* vol. 12, p. 287.
45. Proctor and Gamble, Co.

chapter 15

TEXTILE EMULSIONS

Formula No. 1[1]

Batavia Dammar A/D	10
Stearic Acid	10
Morpholine	1
Water	100

Melt the Dammar and stearic acid. Keeping the temperature at 90-95°C, stir in the morpholine and the water which has been preheated to 90-95°C.

No. 2[2]

White Beef Tallow	30
Sulfonated Tallow (50%)	30
Japan Wax Emulsion	6-8
Water	32-34
Disinfectant or Deodorant	1-2

Melt the Japan wax and sulfonated tallow while agitating; when thoroughly melted, add the beef tallow and stir until thoroughly mixed. Then add water gradually and agitate until a full, white, creamy mix is secured.

No. 3[3]

Rosin	24-60
Linseed Oil	24-60
Alkaline Casein Dispersion	8-10
Sodium Silicate	5-10
Water	500

No. 4[4]

Water	59.6
Beeswax	9.4
Japan Wax	11.0
Stearic Acid	11.0
Oleic Acid	3.0
Caustic Soda	3.0
Olive Oil	3.0

No. 5[5]

Linseed Oil	100
Trichlorethylene	100
Ammonium Linoleate	16
Water	100-200

No. 6[6]

Paraffin Wax	40
Oleic Acid	5
Trigamine	3
Water	112

The trigamine is dissolved in the water and the oleic acid added, stirring thoroughly until completely homogeneous. The mixture is then heated to 65-70°C and the wax, previously melted, added with rapid agitation. Stirring is continued until the emulsion is cold.

No. 7[7]

Melt together at a temperature of 50-60°C equal parts of olein and dried soya bean phosphatides obtained by extraction with 90 parts of benzol and 10 parts of 96% alcohol. By adding an equal quantity of water, keeping the temperature at 50-60°C, and adding small quantities of dilute soda lye, the mixture is emulsified under vigorous stirring. When required for use, this emulsion is diluted with further quantities of water in the desired manner.

No. 8[8]

Rosin	10-15
Olein	2-10
Triethanolamine	7-20
Stearin	7-20
Water	1000

Sizing is done at 50°C for 15 minutes; the temperature is then raised to 60°C. Desizing or washing is done with 0.5% soap solution at 60°C for 1 hour.

No. 9[9]

Sulfonated Castor Oil (58%)	50
Paraffin Oil (28°Bé)	10
Caustic Soda Solution (27°Bé)	12
Water	23
Steam-distilled Pine Oil	12

Mix the ingredients at 35-40°C in the order given, being careful to add the caustic soda solution and pine oil very slowly with constant stirring and allowing the mixture to cool to room temperature as the pine oil is being added.

No. 10[10]

Antistatic Textile Dressing

A. Magnesium Chloride (10% solution)	7
Water	50
B. Mineral Oil	80
Sulfonated Lauryl Alcohol	25
Lauryl Alcohol	5
Oleic Acid	7
Water	50

Add the previously prepared A to B. The dressing is used on cellulose-acetate staple fiber to the extent of 1-4%.

No. 11[11]

Kier Penetrant Oil

Xylol	10
Water	20
Sulfonated Castor Oil (62% T.F.M.)	20

No. 12[12]

Luster Emulsion for Starching

Stearic Acid	1 kg
White Beeswax	200 g
Borax	67-75 kg
Water	4 kg
Citronella Oil for perfume	5 g

10 spoonfuls of this emulsion are used for 1 kg of starch.

No. 13[13]

Rotproofing Emulsion

A. Zinc Naphthenate	60.0

B. Mineral Spirits	17.5
C. Emulsifying Agent	2.5
D. Water	20.0

Dissolve A in B and add slowly with vigorous mixing to C and D (mixed); mix until emulsified. The emulsion is diluted with water before use as desired.

No. 14[14]

Textile-Scouring Assistant

Dupanol LS Paste	15
Hexalin	2
Pine Oil	3
Tetralin	70
Water	10

No. 15[15]

Scouring Agent and Dye Assistant

Condensate NP	18.0-20.0
Water	80.0-82.0

No. 16[16]

Degreasing Emulsion

Kerosene	60
Dupanol LS Paste	5
Water	35

No. 17[17]

Soapy Fulling Oil

A. Oleic Acid	70
Methyl Cyclohexanol	5
Carbon Tetrachloride	20
B. Caustic Potash (50°Bé)	

Add enough B to A to just neutralize to phenolphthalein.

Textile Softeners

No. 18[18]

Water	7
Soap	3
Tallow	1

No. 19[19]

Tallow	100
Caustic Potash (45%)	30

No. 20[20]

Lecithin	0.5- 1.0
Cyclohexanol	5.0- 10.0
Tetralin	50.0-100.0
Water	100.0

No. 21[21]

Mineral Oil	10
Glue, hide	10
Dupanol WS	2
Water	78

Textile Gloss Oil

No. 22[22]

Paraffin Wax	20
Lanette Wax SX	10
Water	70

Heat together and stir until dispersed.

No. 23[23]

Paraffin Wax	20
Petrolatum, white	2
Glycosterin	10
Water	180

Heat together and stir until dispersed.

No. 24[24]

Yarn Finish

Potash Carbonate	2
Rosin	7
Paraffin Wax, soft	10
Montan Wax, bleached, A	10
Water	65

No. 25[25]

Montan Wax, double-bleached	8

Paraffin Wax (40/42°C)	10	Potassium Carbonate	2
Rosin	7	Water	60

Soluble Textile Oils

No. 26, 27

	26[26]	27[27]
Mineral Oil (55 sec Saybolt)	70	82
Oleic Acid	15	3
Sodium Petroleum Sulfonates	15	5
Lard Oil	- -	10

No. 28[28]

Xylol or Toluol	15
Paraffin Oil (28°Bé)	78
Red Oil	2
Alcohol	3
Caustic Soda Solution (27°Bé)	1
Water	1

Mix the paraffin oil and red oil and heat to 40°C. Add the previously mixed water and caustic solution, the xylol slowly, and the alcohol last. When the mixture is uniform, cool as quickly as possible.

No. 29[29]

Water	2.0
Caustic Potash	1.2
Light Mineral Oil	85.0
Diethylene Glycol	2.0
Oleic Acid	9.9

To the mineral oil add 6 parts of oleic acid and the caustic potash dissolved in the water; stir in the diethylene glycol. Then add slowly with constant stirring the remainder of the oleic acid, until the solution becomes clear. If properly made, this oil will emulsify when mixed with water.

No. 30[30]

A.	Turkey Red Oil (50%)	30
	Diglycol Laurate	1
	Alcohol	1
	Methyl Glycol	1
B.	Oleic Acid, white	15
C.	Caustic Potash	to neutralize
	Water	to make 100

No. 31[31]

Of an oil such as a mineral oil, 64 parts are used in admixture with 2 parts of Carbitol, 14 parts of corn oil soap, 10 parts of rosin, 6 parts of water, and 4 parts of diethylene glycol.

No. 32[32]

A.	Oleic Acid	8.0
	Colza Oil	4.0
	Rosin	2.0
	Castor Oil Fatty Acid	2.0
	Spindle Oil	32.0
B.	Alcohol	2.0
	Methyl Glycol	1.0
C.	Caustic Potash (50°)	3.8

No. 33[33]
Textile-Scouring Emulsion

Oleic Acid	10
Pine Oil	87
Trigamine	3
Water	100

Mix 30 parts of the pine oil with the trigamine and oleic acid and stir thoroughly. Then add slowly 35 parts of water and stir till a smooth thick emulsion is formed. Add the remainder of the oil with high-speed agitation and finally the rest of the water.

No. 34[34]
Nonslippery Belt and Rope Preservative

A.	Tallow	6
	Beeswax	1
	Soap	3
B.	Water (boiling)	30

Melt A together and add B slowly with vigorous stirring. When emulsification is complete, dilute with water to 0.950 specific gravity.

No. 35[35]
Spinning Oil for Light or White Fibers

A.	White Oleic Acid	41.0
	Denatured Alcohol	31.0
	Caustic Potash (48° Bé)	17.5
B.	Water	11.0
	White Mineral Oil	854.0
	White Oleic Acid	35.0

Mix A thoroughly and test for neutrality. Make exactly neutral, then add B.

No. 36[36]
Rope Preservative

Tallow	6
Beeswax	1
Water	30
Soap	3

Textile Lubricants

No. 37[37]

Olive Oil	87
Oleic Acid	10
Trigamine	3
Water	100

No. 38[38]

Oleic Acid	40.00
Triethanolamine	2.00
Water	60.00

Stir together at about 35°C and transfer immediately into a homogenizer, where the emulsion is homogenized under about 150 atmospheres pressure.

No. 39[39]

A.	Monoamylamine Oleate	4
	Paraffin Oil (28°Bé)	96
B.	Water	100

Mix A and stir into B.

No. 40[40]
Synthetic Thread Lubricant*

Petrolatum	12.5
Butyl Stearate	9.5
Oleic Acid	8.0
Triethanolamine	2.5
Teflon 30	4.0
Water	63.5

* US Patent 2,870,045

No. 41[41]
Oil Coating for Glass Fibers

Bright Stock (Petroleum) Oil 20-40
Oleic Acid 12.6-18.5
Ammonia (28°) 4
Water to make 100

No. 42[42]
Acetate Rayon Oil

Sulfonated Castor Oil (65%) 50
Commercial Olive Oil 45
Acetic Acid (28%) 20
Paraffin Oil (28°Bé) 5
Water 100

Mix the three oils and the water at 40°C, cool to 30°C, and stir the acetic acid the into mixture slowly.

No. 43[43]
Waterproofing (Acetate) Rayon

Paraffin Wax 5 g
Soap 5 g
Ammonium Hydroxide (0.880) 30 cc
Water 970 cc

Heat to 55°C and stir vigorously until a uniform emulsion results. Soak the rayon in this emulsion at 45°C for 15 minutes, remove; centrifuge and treat for 10 minutes at 20°C with a 5° Tw solution of aluminum acetate.

No. 44[44]
Rayon Delustering

Olive Oil Soap (low titer) 10
White Glue 10
Water 100

Dissolve by warming and stirring; add the paraffin wax. Heat and stir until emulsified. Take $2\frac{1}{2}$ parts of above emulsion and add to it:

Water 100

Stir and add

Infusorial Earth 2.27-4.54

No. 45[45]
Mothproofing Emulsion

A. Santochlor 16
Stearic Acid 1
Carbon Tetrachloride 4
B. Carbowax 4000 6
Triethanolamine 3
Water 5

Solution A, prepared by mixing Santochlor, stearic acid, and carbon tetrachloride, is added with rapid agitation to solution B, prepared by dessolving the Carbowax and triethanolamine in water.

No. 46[46]
Antiseptic Textile

G-4 Technical (Hexachlorophene) 25 lb
Pine Oil 15 gal
Water (180°F) 15 gal
Nekal AEMA 10 lb

Dissolve the G-4 in the pine oil and add slowly with constant stirring to the Nekal water solution.

No. 47[46]

G-4 Technical (Hexachlorophene) 10.0 lb
Isopropyl Alcohol 10.0 lb

Caustic Soda Flakes	1.4 lb
Igepal CA Extra conc.	4.0 lb
Water	9.0 gal

While the use of these emulsions is satisfactory for some purposes, it is not recommended where water leaching is involved, since, they do not contain water repellents.

No. 48[47]

Flash Aging Emulsion

Modified Locust Bean Gum	3 lb
Starch Ether Gum	3 lb
Sodium Lignin Sulfonate	$\frac{3}{4}$ lb
Water	16 gal
Aerotex Resin P116	12 oz
Xylol	1 gal
Mineral Spirits	21 gal

No. 49[47]

Aniline Black Emulsion

Water	24 gal
Sodium Chlorate	12½ lb
Yellow Prussiate of Soda	25 lb
Octyl Alcohol	1 pt
Modified Locust Bean Gum	3 lb
Marasperse N	7 lb
Mineral Spirits	15 gal

No. 50[48]

Azoic Printing Emulsion

Carboxymethyl Cellulose HV 120 or Keltex S, 4% solution (stabilizer)	18 gal
Deceresol Wetting Agent No. 18	3 pt
Caustic Soda, 25% solution	9 pt
Mineral Spirits	20 gal
Oleic Acid	9 pt

No. 51[49]

Screen-Printing Emulsion for Cellulose Acetate*

Kerosene	64.00
Water	26.00
Sodium Lignin Sulfonate	5.00
Glycerol	2.50
Glacial Acetic Acid	2.50
2,4-dinitro-3-hydroxy diphenylamine	0.01

* British Patent 647,098

No. 52[50]

Cotton Piece-Good Dyeings with Phthalocyanine/Solvent System

Solvent L Method (oz/gal)	Total, oz
Phthalogen Brilliant Blue IF3GM	1.00
Phthalogen Brilliant Green IFFB	0.50
Phthalogen K	0.125
Methanol	1.50
Solvent L	14.25
Emulsifier WA	4.22
Urea	14.25
Water and Ice	*q.s.*
	1 gal

No. 53[50]

After-Treatment of Naphtol Dyeings on Continuous Machines

Veropal C Ex conc.	0.5 oz/gal
Emulsifier WA	0.5 oz/gal
Soda Ash Calc	0.5 oz/gal
Water	to 1 gal

Heat and boil for 3—10 sec.

No. 54[51]

Mineral-Oil Emulsion

Mineral-oil emulsions are suitable for use as oil dressings or lubricants in certain textile applications.

White Mineral Oil	43.0
Stearic Acid	5.1
Amino-methyl-propanediol	1.9
Water	50.0

Melt the acid in the oil at about 65°C. Dissolve the amino-methyl-propanediol in the water, also at 65°C. Pour the oil–acid mixture into the water solution with agitation and stir the emulsion occasionally while it cools.

Silicone Water Repellants

No. 55[52]

Dow Corning 1107	18.0
Dow Corning 200 Fluid, 350-12,500 cS.	12.0
Emulphogene BC-720 (610, 720, 840)	2.0
Triton X-100 (Tergitol TMN-6, 10)	0.1
Water	67.9

Mix the Dow Corning silicone fluids and emulsifiers together. Add water with agitation. For maximum stability use a colloid mill homogenizer, etc. This formulation is designed for padding or spraying. If an exhaustive treatment is desired add an acidic cationic catalyst (e.g. Dow Corning XY43).

No. 56[52]

(Protective Colloid System)

Dow Corning 1107	18.0
Dow Corning 200 Fluid, 350-12,500 cS.	12.0
Elvanol 51-05 or 52-22	2.0
Triton X-100	0.2
Water	67.8

Add water to the mixture of the other ingredients with good agitation. See also the preceding formulation.

Softeners

No. 57[52]

Silicone Textile Softener

Dow Corning 200 Fluid, 1000 cs.	40.0
Tergitol TMN-6	2.0
Triton W-30	0.1
Water	57.9

Add water to other the ingredients slowly and with constant agitation. The use of high-shear dispersing equipment will produce a more stable emulsion.

No. 58[53]

Nonionic Softener

Conco HC-100	20.0
Water	80.0

Designed for easy compounding this requires only hot water and agitation. The finished product shows good resistance to yellowing and produces a very soft, full hand.

No. 59[53]

Cationic Softener

Conco SC-100	20.0
Glacial Acetic Acid	7.0

Water 73.0

Melt the Conco SC-100 at 180°F, add the glacial acetic acid, and then water. Cool and drain at 120°F. The product is recommended where good nonyellowing characteristics are required.

References

1. H. Bennett, *Chemical Formulary*, vol. 5, p. 591, Chemical Publishing, New York, 1941.
2. Personal Communication.
3. Lefranc, US Patent 1 861 927 (1932).
4. Bennett, *op. cit.* vol. 4, p. 529, 1939.
5. *Ibid.* vol. 1, p. 475, 1933.
6. *Ibid.* vol. 4, p. 553, 1939.
7. Personal Communication.
8. Bennett, *op. cit.* vol. 2, p. 468, 1935.
9. *Ibid.* vol. 3, p. 365, 1936.
10. *Ibid.* vol. 5, p. 595, 1941.
11. *Ibid.* vol. 3, p. 365, 1936.
12. *Ibid.* vol. 5, p. 596, 1941.
13. *Ibid.* p. 593, 1941.
14. E. I. du Pont de Nemours and Co., Wilmington, Del.
15. Continental Chemical Co.
16. E. I. du Pont de Nemours and Co., Wilmington, Del.
17. Bennett, *op. cit.* vol. 4, p. 527, 1939.
18. *Ibid.* vol. 4, p. 526, 1939.
19. *Ibid.*
20. *Ibid.*
21. E. I. du Pont de Nemours and Co., Wilmington, Del.
22. Bennett, *op. cit.* vol. 2, p. 455, 1935.
23. *Ibid.*
24. *Ibid.* vol. 5, p. 596, 1941.
25. *Ibid.* vol. 4, p. 188, 1939.
26. Kinney, Can adian Patent 284,807 (1928).
27. *Ibid.*
28. Bennett, *op. cit.* vol. 3, p. 365, 1936.
29. *Ibid.* vol. 4, p. 188, 1939.
30. *Ibid.*
31. *Ibid.* vol. 3, p. 365, 1936.
32. *Ibid.* vol. 4, p. 187, 1939.
33. *Ibid.* p. 499, 1939.
34. *Ibid.* vol. 5, p. 593, 1941.
35. *Ibid.* vol. 4, p. 188, 1939.
36. Pfahl, British Patent 488,643 (1938).
37. Bennett, *op. cit.* vol. 4, p. 187, 1939.
38. *Ibid.*

39. *Ibid*.
40. *Ibid*. vol. 11, p. 243.
41. *Ibid*. vol. 4, p. 215, 1939.
42. *Ibid*. vol. 3, p. 364, 1936.
43. *Ibid*. vol. 2, p. 468, 1935.
44. *Ibid*.
45. *Ibid*. vol. 7, p. 402, 1945.
46. *Ibid*. vol. 11, p. 353.
47. *Ibid*. p. 345.
48. *Ibid*. p. 344.
49. *Ibid*. p. 346.
50. Verona-Pharma Chemical Corp.
51. Commercial Solvents Corp.
52. Dow Corning Corp.
53. Continental Chemical Co.

APPENDIX

CONVERSION TABLES

Length

1m = 1000 mm = 3.28 ft = 39.37 in.
1 ft = 304.8 mm
1 in. = 25.4 mm

Area

1 m^2 = 10.76 ft^2 = 1550 $in.^2$
1 ft^2 = 929 cm^2
1 $in.^2$ = 6.45 cm^2

Volume

1 m^3 = 1000 l = 6102 $in.^3$ = 264 gal (l = liter; gal = US gallons)
1 gal = 3785 cm^3
1 ft^3 = 28.3 l = 7.48 gal

Flow

1 m^3/hr = 4.4 gal/min = 0.0098 ft^3/sec
1 l/sec = 15.9 gal/min = 3.6 m^3/hr
1 gal/min = 0.227 m^3/hr = 0.063 l/sec

Mass

1 ton = 1000 kg = 2205 lb
1 lb = 454 g
1 kg = 2.2 lb
1 g = 0.0353 oz

Density

1 g/cm^3 = 1kg/dm^3 = 1t/m^3 = 62.43 lb/ft^3 (t = ton = 1000 kg)
1 lb/ft^3 = 0.016 kg/dm^3
Baumé scale: kg/dm^3 = 145/(145 − deg. Bé)

Pressure

1 at = 1 kg/cm^2 = 10,000 kg/m^2 = 10 m water column at 4°C = 735 torr = 14.2 psi (at = atmosphere; torr = mm Hg at 0°C)
1 psi = 0.0703 kg/cm^2

Viscosity

Dynamic: 1 poise = 1 dyne-second/cm^2
Kinematic: 1 stoke = 1 cm/sec^2 (= viscosity/density)

Heat

1 kcal=3.968 btu
1 kcal/kg=1.8 btu/lb
1 kcal/l=15.015 btu/gal
1 kcal/m^2=0.00256 btu/in.^{2}
1 btu=0.252 kcal
1 btu/lb=0.556 kcal/kg
1 btu/in.^{2}=390.6 kcal/m^2
1 btu/gal=0.0666 kcal/l

Power

1 kW=1.34 hp=102 kg/sec=737.54 ft-lb/sec=0.948 btu/sec=0.2388 kcal/sec
1 hp=0.746 kW=76.1 kg/sec=550 ft-lb/sec=0.707 btu/sec=0.178 kcal/sec

Work

1 kWhr=860 kcal=3412 btu=2.66×10^6 ft-lb=367,098 kg

Table of Unit prefixes (National Bureau of Standards)

Multiples and submultiples		*Prefixes*	*Symbols*
1,000,000,000,000	= 10^{12}	tera-	T
1,000,000,000	= 10^9	giga-	G
1,000,000	= 10^6	mega-	M
1,000	= 10^3	kilo-	k
100	= 10^2	hecto-	h
10	= 10	deka-	D
.1	= 10^{-1}	deci-	d
.01	= 10^{-2}	centi-	c
.001	= 10^{-3}	milli-	m
.000001	= 10^{-6}	micro-	μ
.000000001	= 10^{-9}	nano-	n
.000000000001	= 10^{-12}	pico-	p

VISCOSITY COMPARISON CHART

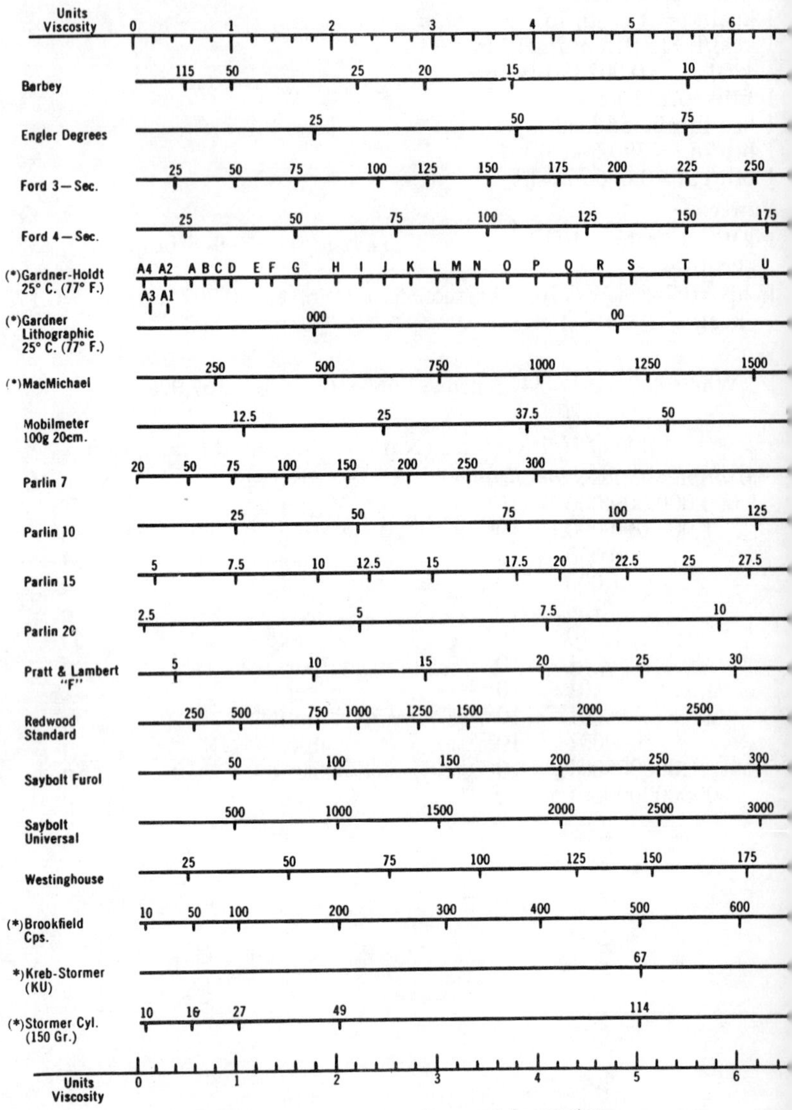

Use "Units Viscosity" scales top and bottom as reference marks for comparisons.

Scales marked (*) may be compared with each other or read directly to "Poises" (Absolute Viscosity) on
Viscosity Scale.

On all other scales direct conversions may be made between scales or read directly to "Stokes" (Kinematic
on the Units Viscosity Scale.

To convert Kinematic to Absolute Viscosity multiply "Stokes" by density to obtain "Poises." Conversely, to

VISCOSITY COMPARISON CHART (*continued*)

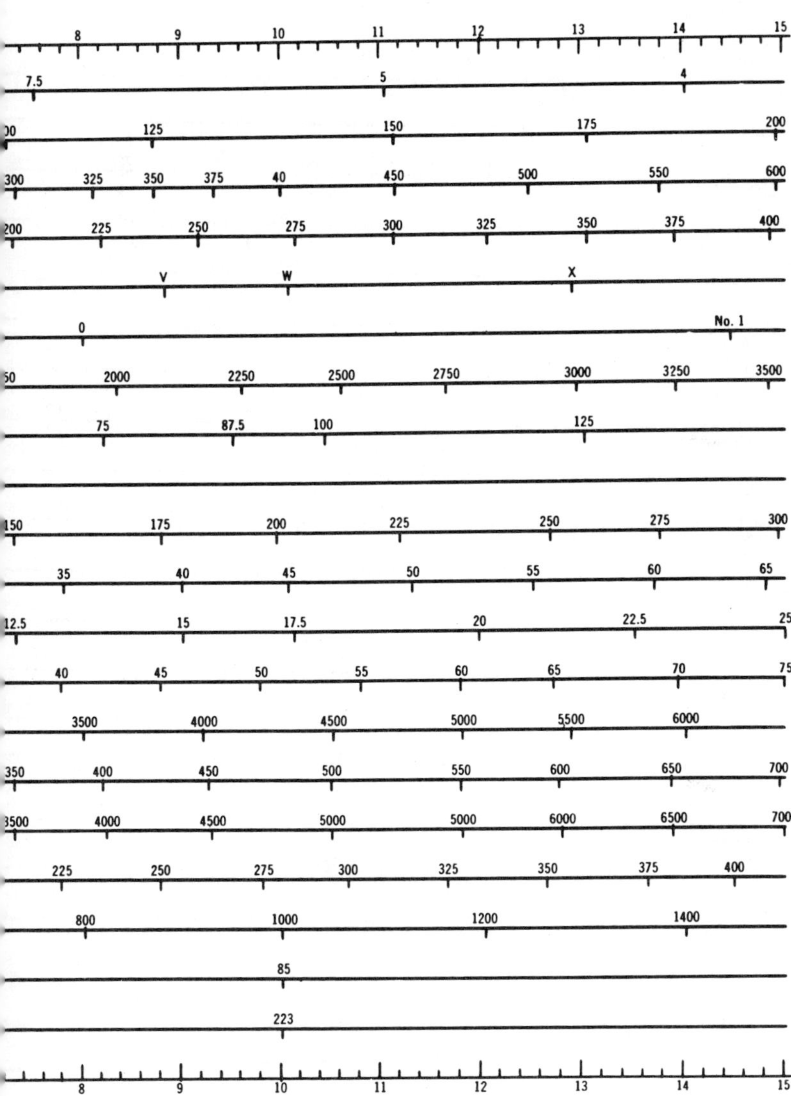

Absolute to Kinematic Viscosity divide "Poises" by density to obtain "Stokes."

Kinematic Viscosity is the ratio of Absolute Viscosity to density ($d\frac{+}{4}$).

With the exception of Gardner-Holdt and Gardner Lithographic Scales comparisons are made directly at the same temperatures.

VISCOSITY COMPARISON CHART (*continued*)

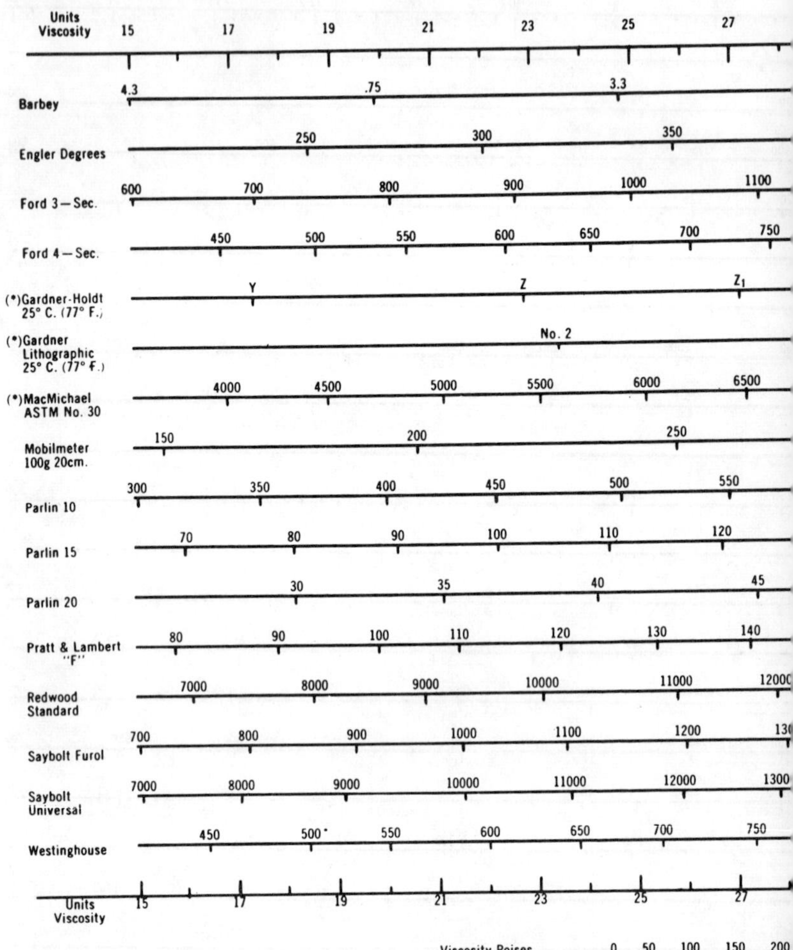

Viscosity Poises	0	50	100	150	200
Gardner-Holdt 25° C. (77° F.)	Z_3 Z_4		Z_5	Z_6	
Gardner Lithographic 25° C. (77° F.)		No. 4		No. 5	No.

VISCOSITY COMPARISON CHART (*continued*)

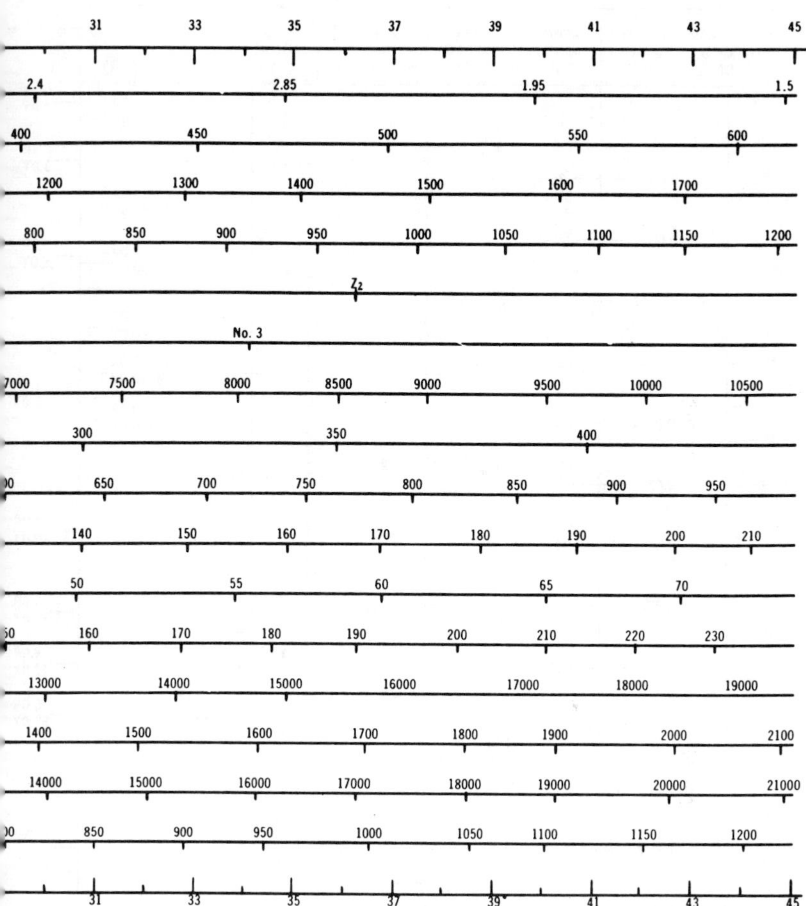

(AMPLE

convert Saybolt Universal, 1000 Seconds, 100°C (212°F) density 0.800 to Gardner-Holdt. Read from Saybolt Scale Units Viscosity Scale = 2.0 Stokes (Kinematic). Multiply by density. 0.800 = 1.6 Poises (Absolute). Read from 1.6 Units cosity Scale to Gardner-Holdt Scale = G. Express Viscosity as Gardner-Holdt "G." In other words, this material at)°C has the same Viscosity that the Gardner-Holdt "G" Standard has at 25°C.

convert Saybolt Universal, 1000 seconds, 212°F, to Engler de-es. Read with straight-edge lined up with Saybolt Universal 0 and reference marks top and bottom to Engler 27 degrees ° F.

COLOR STANDARDS COMPARATOR

(Harchem Div., Wallace & Tiernan Inc., Bellevillle, N.J.)

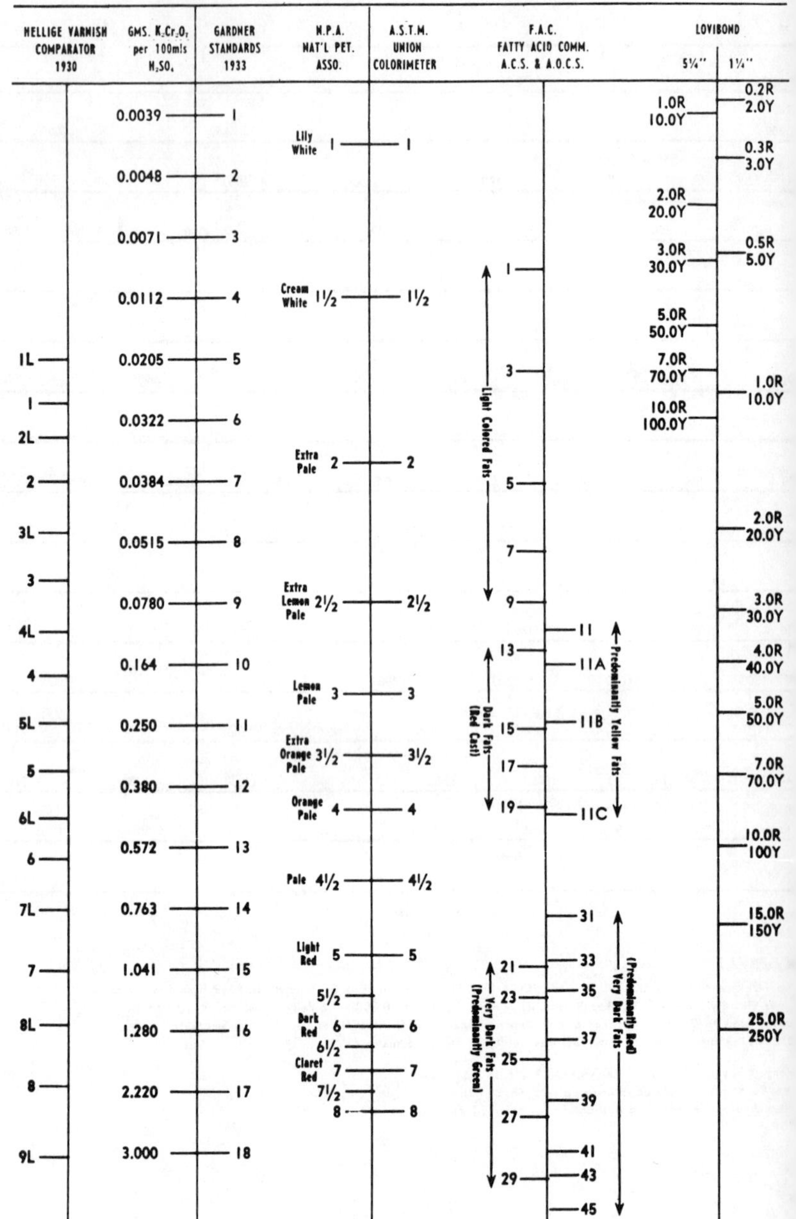

C ←	F / C	→ F
-40.00	-40	-40.0
-39.44	-39	-38.2
-38.89	-38	-36.4
-38.33	-37	-34.6
-37.78	-36	-32.8
-37.22	-35	-31.0
-36.67	-34	-29.2
-36.11	-33	-27.4
-35.56	-32	-25.6
-35.00	-31	-23.8
-34.44	-30	-22.0
-33.89	-29	-20.2
-33.33	-28	-18.4
-32.78	-27	-16.6
-32.22	-26	-14.8
-31.67	-25	-13.0
-31.11	-24	-11.2
-30.56	-23	-9.4
-30.00	-22	-7.6
-29.44	-21	-5.8
-28.89	-20	-4.0
-28.33	-19	-2.2
-27.78	-18	-0.4
-27.22	-17	1.4
-26.67	-16	3.2
-26.11	-15	5.0
-25.56	-14	6.8
-25.00	-13	8.6
-24.44	-12	10.4
-23.89	-11	12.2
-23.33	-10	14.0
-22.78	-9	15.8
-22.22	-8	17.6
-21.67	-7	19.4
-21.11	-6	21.2
-20.56	-5	23.0
-20.00	-4	24.8
-19.44	-3	26.6
-18.89	-2	28.4
-18.33	-1	30.2
-17.78	0	32.0
-17.22	1	33.8
-16.67	2	35.6
-16.11	3	37.4
-15.56	4	39.2

C ←	F / C	→ F
-15.00	5	41.0
-14.44	6	42.8
-13.89	7	44.6
-13.33	8	46.4
-12.78	9	48.2
-12.22	10	50.0
-11.67	11	51.8
-11.11	12	53.6
-10.56	13	55.4
-10.00	14	57.2
-9.44	15	59.0
-8.89	16	60.8
-8.33	17	62.6
-7.73	18	64.4
-7.22	19	66.2
-6.67	20	68.0
-6.11	21	69.8
-5.56	22	71.6
-5.00	23	73.4
-4.44	24	75.2
-3.89	25	77.0
-3.33	26	78.8
-2.78	27	80.6
-2.22	28	82.4
-1.67	29	84.2
-1.11	30	86.0
-0.56	31	87.8
0.00	32	89.6
0.56	33	91.4
1.11	34	93.2
1.67	35	95.0
2.22	36	96.8
2.78	37	98.6
3.33	38	100.4
3.89	39	102.2
4.44	40	104.0
5.00	41	105.8
5.56	42	107.6
6.11	43	109.4
6.67	44	111.2
7.22	45	113.0
7.78	46	114.8
8.33	47	116.6
8.89	48	118.4
9.44	49	120.2

C ←	F / C	→ F
10.00	50	122.0
10.56	51	123.8
11.11	52	125.6
11.67	53	127.4
12.22	54	129.2
12.78	55	131.0
13.33	56	132.8
13.89	57	134.6
14.44	58	136.4
15.00	59	138.2
15.56	60	140.0
16.11	61	141.8
16.67	62	143.6
17.22	63	145.4
17.78	64	147.2
18.33	65	149.0
18.89	66	150.8
19.44	67	152.6
20.00	68	154.4
20.56	69	156.2
21.11	70	158.0
21.67	71	159.8
22.22	72	161.6
22.78	73	163.4
23.33	74	165.2
23.89	75	167.0
24.44	76	168.8
25.00	77	170.6
25.56	78	172.4
26.11	79	174.2
26.67	80	176.0
27.22	81	177.8
27.78	82	179.6
28.33	83	181.4
28.89	84	183.2
29.44	85	185.0
30.00	86	186.8
30.56	87	188.6
31.11	88	190.4
31.67	89	192.2
32.22	90	194.0
32.78	91	195.8
33.33	92	197.6
33.89	93	199.4
34.44	94	201.2

C ←	F / C	→ F
35.00	95	203.0
35.56	96	204.8
36.11	97	206.6
36.67	98	208.4
37.22	99	210.2
37.78	100	212.0
38.33	101	213.8
38.89	102	215.6
39.44	103	217.4
40.00	104	219.2
40.56	105	221.0
41.11	106	222.8
41.67	107	224.6
42.22	108	226.4
42.78	109	228.2
43.33	110	230.0
43.89	111	231.8
44.44	112	233.6
45.00	113	235.4
45.56	114	237.2
46.11	115	239.0
46.67	116	240.8
47.22	117	242.6
47.78	118	244.4
48.33	119	246.2
48.89	120	248.0
49.44	121	249.8
50.00	122	251.6
50.56	123	253.4
51.11	124	255.2
51.67	125	257.0
52.22	126	258.8
52.78	127	260.6
53.33	128	262.4
53.89	129	264.2
54.44	130	266.0
55.00	131	267.8
55.56	132	269.6
56.11	133	271.4
56.67	134	273.2
57.22	135	275.0
57.78	136	276.8
58.33	137	278.6
58.89	138	280.4
59.44	139	282.2

C ←	F / C	→ F
60.00	140	284.0
65.56	150	302.0
71.11	160	320.0
76.67	170	338.0
82.22	180	356.0
87.78	190	374.0
93.33	200	392.0
98.89	210	410.0
104.44	220	428.0
110.00	230	446.0
115.56	240	464.0
121.11	250	482.0
126.67	260	500.0
132.22	270	518.0
137.78	280	536.0
143.33	290	554.0
148.89	300	572.0
154.44	310	590.0
160.00	320	608.0
165.56	330	626.0
171.11	340	644.0
176.67	350	662.0
182.22	360	680.0
187.78	370	698.0
193.33	380	716.0
198.89	390	734.0
204.44	400	752.0

PROPORTIONAL PARTS

C ← ADD	F / C	→ F ADD
0.55	1	1.8
1.11	2	3.6
1.66	3	5.4
2.22	4	7.2
2.77	5	9.0
3.33	6	10.8
3.88	7	12.6
4.44	8	14.3
5.00	9	16.2
5.55	10	18.0

For .1 degree divide by 10.
Thus 16.2°C = (60.8 + .36)°F

RATE OF FLOW CHART

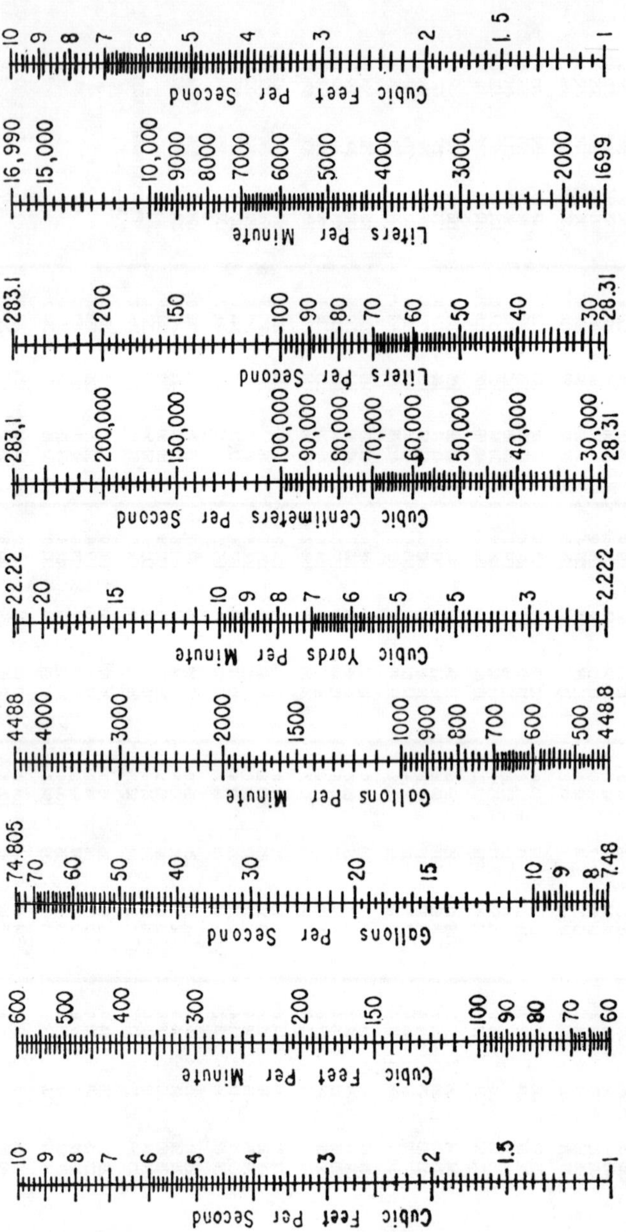

List Of Emulsifying Agents

Name	Supplier	Anionic					Nonionic				Cationic			Natural					
		Sulfates	Alkyl Sulfonic Acids and Their Salts	Carboxylic Acids and Their Salts	Alkylaryl Sulfonic Acids and Salts	Miscellaneous	Polyalkoxy Ethers	Alkoxy Esters	Alkanolamides	Miscellaneous	Amines and Amine Salts	Quaternary Amines and Their Salts	Miscellaneous	Amphoteric	Cellulose Derivatives	Water Soluble Gums	Lipids	Fluorocarbons	Silanes and Silicones
Abex	Alcolac Chemical Corp.					×													
Actiflo 68, 70	Central Soya Co.																×		
Acto	Humble Oil and Refining Co.				×														
Ahcol	I.C.I. Organics, Inc.	×																	
Ahcovel FP	I.C.I. Organics, Inc.							×											
Ahcowet ANS	I.C.I. Organics, Inc.				×														
Ahcowet RS	I.C.I. Organics, Inc.	×																	
Ahcowet SDS	I.C.I. Organics, Inc.		×																
Alacsan	Alcolac Chemical Corp.											×							
Alcolan	Robinson Wagner Co., Inc.							×											
Aldo	Glyco Chemicals, Inc.							×											
Aldosperse	Glyco Chemicals, Inc.							×											
Alkapent	Wayland Chemical Corp.					×													
Aliquat	General Mills											×							
Amber	Proctor and Gamble Co.			×															
Amerchol	American Cholesterol Products, Inc.							×											

Name	Supplier	Anionic					Nonionic				Cationic			Natural					
		Sulfates	Alkyl Sulfonic Acids and Their Salts	Carboxylic Acids and Their Salts	Alkylaryl Sulfonic Acids and Salts	Miscellaneous	Polyalkoxy Ethers	Alkoxy Esters	Alkanolamides	Miscellaneous	Amines and Amines Salts	Quaternary Amines and Their Salts	Miscellaneous	Amphoteric	Cellulose Derivatives	Water Soluble Gums	Lipids	Fluorocarbons	Silanes and Silicones
Amerlate	American Cholesterol Products, Inc.							×											
Amphocerin	Henkel International GmbH, A. H. Carnes							×											
Aquarex G	E.I. duPont deNemours & Co., Inc.		×																
Araphen	Henkel International GmbH, A.H. Carnes										×								
Arizona DRS-30, 32, 33	Arizona Chemical Co.			×															
Arlacel*	Atlas Chemical Industries, Inc.							×											
Atcor HC	Atlas Chemical Industries, Inc.										×								
Atlox	Atlas Chemical Industries, Inc.							×											
Atmos V50, 300	Atlas Chemical Industries, Inc.							×											
Atmul 80-651	Atlas Chemical Industries, Inc.							×											
Atpet 80-200	Atlas Chemical Industries, Inc.							×											
Benax 2A1	The Dow Chemical Co.				×														
Brij 30-98	Atlas Chemical Industries, Inc.						×												
Busopal NA	BASF Colors and Chemicals, Inc.				×														
Calcium Sulfonate	Swift & Co.				×														
Calimulse PRS, DMS	Pilot Chemical Co.		×																
Calsoft F90, L60	Pilot Chemical Co.				×														
Carbopol	B.F. Goodrich Chemical Co.					×													
Carolate	Robeco Chemicals, Inc.								×										
CC-16	National Lead Co.					×													

* Series of similar products.

Name	Supplier	Anionic					Nonionic				Cationic			Natural					
		Sulfates	Alkyl Sulfonic Acids and Their Salts	Carboxylic Acids and Their Salts	Alkylaryl Sulfonic Acids and Salts	Miscellaneous	Polyalkoxy Ethers	Alkoxy Ethers	Alkanolamides	Miscellaneous	Amines and Amine Salts	Quaternary Amines and Their Salts	Miscellaneous	Amphoteric	Cellulose Derivatives	Water Soluble Gums	Lipids	Fluorocarbons	Silanes and Silicones
Centrol 1P, 1F, 2F, 3P, 3F, 5F	Central Soya Co.																×		
Centrolene	Central Soya Co.																×		
Centrolex	Central Soya Co.																×		
Centromix	Central Soya Co.																×		
Centrophil	Central Soya Co.																×		
Ceralan	Robinson Wagner Co., Inc.									×									
Cetrimide	I.C.I. Organics, Inc.											×							
Chemactant A-5, LM-14	American Cholesterol Products, Inc.									×									
Chemactant AR	American Cholesterol Products, Inc.			×															
Chemactant Q	American Cholesterol Products, Inc.											×							
Chemactant S	American Cholesterol Products, Inc.						×												
Chemadene 300	Richardson Co.													×					
Concental 75	I.C.I. Organics, Inc.	×																	
Conco AAS Special # 3	Continental Chemical Co.				×														
Conco Emulsifier K	Continental Chemical Co.								×										
Conco NI 21-NI190, NI300	Continental Chemical Co.						×												
Concopal, A, AS, S, SS	Continental Chemical Co.	×																	
Comperlan	Henkel International GmbH, A.H. Carnes								×										
Complexine G	Guardian Chemical Corp.		×																
Cordon 900	Finetex, Inc.	×																	
Crodafos N-3, N-10	Croda, Inc.										×								

Name	Supplier	Anionic					Nonionic				Cationic			Natural					
		Sulfates	Alkyl Sulfonic Acids and Their Salts	Carboxylic Acids and Their Salts	Alkylaryl Sulfonic Acids and Salts	Miscellaneous	Polyalkoxy Ethers	Alkoxy Esters	Alkanolamides	Miscellaneous	Amines and Amine Salts	Quaternary Amines and Their Salts	Miscellaneous	Amphoteric	Cellulose Derivatives	Water Soluble Gums	Lipids	Fluorocarbons	Silanes and Silicones
Crodafos N-3 neutral, N-10 neutral	Croda, Inc.										×								
Dehydag Wax E	Henkel International GmbH, A.H. Carnes	×																	
Dehydazol	Henkel International GmbH, A.H. Carnes														×				
Dehydol	Henkel International GmbH, A.H. Carnes						×												
Dehydrophen	Henkel International GmbH, A.H. Carnes						×												
Dehymin	Henkel International GmbH, A.H. Carnes										×								
Dehymuls	Henkel International GmbH, A.H. Carnes							×											
Dehyquart	Henkel International GmbH, A.H. Carnes											×							
Deriphat 151, 151C, 170C, 154, 160,	General Mills													×					
Detergent 234	Dixo Company, Inc.									×									
Diam	General Mills										×								
Dispersant NI	Chevron Chemical Co., Oronite Division						×												
Dow Corning 113-202	Dow Corning Corporation																		×
Dowfax 9N 4-40	The Dow Chemical Co.						×												
Duponol	E.I. du Pont de Nemours & Co. Inc.	×																	
Ecconol B	Essential Chemicals Corp.								×										
Ecconol 70	Essential Chemicals Corp.							×											
Emulphor A Oil Soluble	BASF Colors and Chemicals, Inc.							×											

Name	Supplier	Anionic					Nonionic				Cationic			Natural					
		Sulfates	Alkyl Sulfonic Acids and Their Salts	Carboxylic Acids and Their Salts	Alkylaryl Sulfonic Acids and Salts	Miscellaneous	Polyalkoxy Ethers	Alkoxy Esters	Alkanolamides	Miscellaneous	Amines and Amine Salts	Quaternary Amines and Their Salts	Miscellaneous	Amphoteric	Cellulose Derivatives	Water Soluble Gums	Lipids	Fluorocarbons	Silanes and Silicones
Emulphor AF, P, O, O Type 1178C	BASF Colors and and Chemicals, Inc.						×												
Emulphor EL, FJ	BASF Colors and Chemicals, Inc.						×												
Emulphor FM Oil Soluble	BASF Colors and Chemicals, Inc.							×											
Emulsifier DL	Finetex, Inc.					×													
Emulsifier M85, K1	Varney Chemical Corp.												×						
Emulsifier WA	Verona-Pharma Chemical Corp.							×											
Estrex 1G 5-58	Swift & Co.			×															
Estrex P4	Swift & Co.							×											
Estrex SE 550, 750, 950	Swift & Co.							×											
Ethosperse	Glyco Chemicals, Inc.						×												
Emulgin	Henkel International GmbH, A.H. Carnes Co.						×												
EZ MUL	National Lead Co.								×										
FC 95, 98, 128, 134, FX 161	3M Company																	×	
G263, G271	Atlas Chemical Industries, Inc.	×																	
G-1045, G-3284, G-1288, G-1292, G-1295	Atlas Chemical Industries, Inc.							×											
G-1441, G-1471	Atlas Chemical Industries, Inc.						×												
G1702, G1726, G2079, G2127 G2162	Atlas Chemical Industries, Inc.							×											
G1790, G1795	Atlas Chemical Industries, Inc.						×												
G3300	Atlas Chemical Industries, Inc.				×														
G3570, G3780A	Atlas Chemical Industries, Inc.										×								
G3634A	Atlas Chemical Industries, Inc.											×							
Glycomul	Glyco Chemicals, Inc.							×											

Name	Supplier	Anionic					Nonionic				Cationic			Natural					
		Sulfates	Alkyl Sulfonic Acids and Their Salts	Carboxylic Acids and Their Salts	Alkylaryl Sulfonic Acids and Salts	Miscellaneous	Polyalkoxy Ethers	Alkoxy Esters	Alkanolamides	Miscellaneous	Amines and Amine Salts	Quaternary Amines and Their Salts	Miscellaneous	Amphoteric	Cellulose Derivatives	Water Soluble Gums	Lipids	Fluorocarbons	Silanes and Silicones
Glycosperse	Glyco Chemicals, Inc.							×											
Hartolan	Croda, Inc.									×									
Hodag 20-152	Hodag Chemical Corp.							×											
Hodag C23, 4, 5, B20	Hodag Chemical Corp.										×								
Hodag C100, 121, 130, C120, 122, 127	Hodag Chemical Corp.												×						
Hodag C128, 132	Hodag Chemical Corp.												×						
Hodag CQ 1, 2, 3, 14	Hodag Chemical Corp.											×							
Hodag DGL, DGO, DGS, EGS, EDGS, MGDT, PGS, PGL, GML, GMO, GMS, GMP, GMT, GTO, GMY, GMF, PGMS, S-35	Hodag Chemical Corp.								×										
Hodag PSML, PSMO, PSTO, PSMP, PSMS	Hodag Chemical Corp.							×											
Hodag SA	Hodag Chemical Corp.								×										
Hodag SML, SMO, STO, SMP, SMS, STS	Hodag Chemical Corp.							×											
Hodag SMO, SVO, SUS, PGMP, PGS, GMS	Hodag Chemical Corp.							×											
Hyamine	Rohm & Haas Co.											×							
ICE # 1	Glidden Co.							×											
Intexsan ABM-50, LB-MB, SB85	Washine Chemical Corp.											×							
Invermul	National Lead Co.					×													

Name	Supplier	Anionic					Nonionic				Cationic			Natural					
		Sulfates	Alkyl Sulfonic Acids and Their Salts	Carboxylic Acids and Their Salts	Alkylaryl Sulfonic Acids and Salts	Miscellaneous	Polyalkoxy Ethers	Alkoxy Esters	Alkanolamides	Miscellaneous	Amines and Amine Salts	Quaternary Amines and Their Salts	Miscellaneous	Amphoteric	Cellulose Derivatives	Water Soluble Gums	Lipids	Fluorocarbons	Silanes and Silicones
Iso-lan	Goldschmidt Chemical Corp.									×									
Isomul Extra	Isochem Corp.								×										
Isoplast M, D	Isochem Corp.							×											
Ivory	Procter and Gamble, Co.			×															
Kamar O	Finetex, Inc.	×																	
Kelecin	Textron, Inc.															×			
Kyro AC	Procter and Gamble, Co.			×															
Kyro EOB	Procter and Gamble, Co.						×												
Lanogel	Robinson Wagner Co., Inc.						×												
Lauryl Pyridinium Chloride	Hooker Chemical Corp.												×						
Lissapol 500, ACA	I.C.I. Organics, Inc.	×																	
Lissolamine	I.C.I. Organics, Inc.											×							
Lubrol	I.C.I. Organics, Inc.						×												
Maravil	Henkel International GmbH, A.H. Carnes Co.				×														
Methocel HG	The Dow Chemical Co.														×				
Methocel MC	The Dow Chemical Co.														×				
Miranol, Miranol 2MCA	Miranol Chemical Co., Inc.													×					
Miranol C2M and JEM Anhy. Acid	Miranol Chemical Co., Inc.													×					
Miranol SHD Conc.	Miranol Chemical Co., Inc.													×					
Monamides	Mona Industries, Inc.								×										
Monamine	Mona Industries, Inc.										×								
Monateric	Mona Industries, Inc.													×					

Name	Supplier	Anionic					Nonionic				Cationic			Natural					
		Sulfates	Alkyl Sulfonic Acids and Their Salts	Carboxylic Acids and Their Salts	Alkylaryl Sulfonic Acids and Salts	Miscellaneous	Polyalkoxy Ethers	Alkoxy Esters	Alkanolamides	Miscellaneous	Amines and Amine Salts	Quaternary Amines and Their Salts	Miscellaneous	Amphoteric	Cellulose Derivatives	Water Soluble Gums	Lipids	Fluorocarbons	Silanes and Silicones
Monazolines	Mona Industries, Inc.												×						
Mulsoid 815 D, M, S	Colloids, Inc.									×									
Myrj 45-53	Atlas Chemical Industries, Inc.							×											
Nacan	Allied Chemical Corp. Nat'l Aniline Div.				×														
Naccolene A Concentrate	Allied Chemical Corp. Nat'l Aniline Div.		×																
Nacconol 40 Series DBX, 355L, 60SL	Allied Chemical Corp. Nat'l, Aniline Div.				×														
Nacconol 60TL, 88SA, 98SA	Allied Chem. Corp., Nat'l. Aniline Div.				×														
Nekal AEM	BASF Colors and Chemicals, Inc.				×														
Nekanil C Highly Conc.	BASF Colors and Chemicals, Inc.						×												
Nekanil LN, 907	BASF Colors and Chemicals, Inc.						×												
Nekanil W Extra	BASF Colors and Chemicals, Inc.								×										
Nonionic E2-E30, 0-8-0-12	Hodag Chemical Corp.						×												
Nonionic L, TD, S, GR	Hodag Chemical Corp.						×												
Non-Pareil	Procter and Gamble, Co.			×															
Olate	Procter and Gammble, Co.			×															
Orvus AB	Procter and Gamble, Co.		×																
Orvus ES, K, WA	Procter and Gamble, Co.	×																	
Pegosperse	Glyco Chemicals, Inc.							×											
Petrosul	Pennsylvania Refining Co.		×																
P & G Amides	Procter and Gamble, Co.								×										
P & G Emulsifier No. 104	Procter and Gamble, Co.	×																	

Name	Supplier	Anionic					Nonionic				Cationic			Natural					
		Sulfates	Alkyl Sulfonic Acids and Their Salts	Carboxylic Acids and Their Salts	Alkylaryl Sulfonic Acids and Salts	Miscellaneous	Polyalkoxy Ethers	Alkoxy Esters	Alkanolamides	Miscellaneous	Amines and Amine Salts	Quaternary Amines and Their Salts	Miscellaneous	Amphoteric	Cellulose Dervatives	Water Soluble Gums	Lipids	Fluorocarbons	Silanes and Silicones
Plurofac	Wyandotte Chemi cals Corp.						×												
Polawax	Croda, Inc.						×												
Polychol 5-40	Croda, Inc.						×												
Promulgen	Robinson Wagner Co., Inc.						×												
Protegin X	Goldschmidt Chemical Corp.									×									
Renex 20	Atlas Chemical Industries, Inc.							×											
Renex 30-36	Atlas Chemical Industries, Inc.						×												
Ricilan	American Choles-terol Products, Inc.							×											
Rueterg 97-S	Finetex, Inc.				×														
Rueterg IPA	Finetex, Inc.	×																	
Sandrol	Clintwood Chemi-cal Co.								×										
Saratoga	Procter and Gam-ble, Co.			×															
SGF	Glidden Co.							×											
Sipenol	Alcolac Chemical Corp.										×								
Sipex	Alcolac Chemical Corp.	×																	
Sipon	Alcolac Chemical Corp.	×																	
Siponate	Alcolac Chemical Corp.				×														
Siponic	Alcolac Chemical Corp.						×												
SL	Glidden Co.									×									
Sohilan	American Choles-terol Products, Inc.						×												
Solan	Croda, Inc.						×												
Soromin HS	BASF Colors and Chemicals, Inc.												×						
Span 20-85	Atlas Chemical Industries, Inc.							×											

Name	Supplier	Anionic					Nonionic				Cationic			Natural					
		Sulfates	Alkyl Sulfonic Acids and Their Salts	Carboxylic Acids and Their Salts	Alkylaryl Sulfonic Acids and Salts	Miscellaneous	Polyalkoxy Ethers	Alkoxy Esters	Alkanolamides	Miscellaneous	Amines and Amine Salts	Quaternary Amines and Their Salts	Miscellaneous	Amphoteric	Cellulose Derivatives	Water Soluble Gums	Lipids	Fluorocarbons	Silanes and Silicones
SPG 187	Glidden Co.							×											
Standamul	Standard Chemical Products						×												
Sul-Fon-ate OA5	Tennessee Corp.		×																
Sulfonated Castor Oil No. 50, 75	Swift & Co.		×																
Sulfonated Pure Neats Foot	Swift & Co.		×																
Sulfonated Soybean Oil	Swift & Co.		×																
Sulfonated Sperm Oil	Swift & Co.		×																
Sulfonated Red Oil	Swift & Co.		×																
Sulfopon	Henkel International GmbH, A.H. Carnes Co.	×																	
Surco 5024	Surfact Co., Inc.								×										
Surco AM-LS, TEA-LS	Surfact Co., Inc.	×																	
Surco Coconut Condensate	Surfact Co., Inc.								×										
Surfactol 318, 340, 365, 380	Baker Castor Oil Co.							×											
Surfonic LF-5, 6, 7, 16, 17	Jefferson Chemical Co.						×												
Surfonic N Series	Jefferson Chemical Co.						×												
Synotex	Dixo Company, Inc.									×									
Synthrapol	I.C.I. Organics, Inc.						×												
T-Det C-30, N	Thompson-Hayward Chem. Co.						×												
Tegacid Regular	Goldschmidt Chemical Corp.							×											
Tegacid Special	Goldschmidt Chemical Corp.			×															
Tegamine	Goldschmidt Chemical Corp.												×						

Name	Supplier	Anionic					Nonionic				Cationic			Natural					
		Sulfates	Alkyl Sulfonic Acids and Their Salts	Carboxylic Acids and Their Salts	Alkylaryl Sulfonic Acids and Salts	Miscellaneous	Polyalkoxy Ethers	Alkoxy Esters	Alkanolamides	Miscellaneous	Amines and Amine Salts	Quaternary Amines and Their Salts	Miscellaneous	Amphoteric	Cellulose Derivatives	Water Soluble Gums	Lipids	Fluorocarbons	Silanes and Silicones
Tegin P	Goldschmidt Chemical Corp.			×															
Tergitol 3A6	Union Carbide Chemicals Co.						×												
Tergitol 08, 4, 7	Union Carbide Chemicals Co.	×																	
Tergitol 12 M 10	Union Carbide Chemicals Co.									×									
Tergitol 12P Series, 15-S Series	Union Carbide Chemicals Co.						×												
Tergitol 15-S-3A, 15-S-3S	Union Carbide Chemicals Co.	×																	
Tergitol NP, TP Series, OP	Union Carbide Chemicals Co.						×												
Tergitol P-28	Union Carbide Chemicals Co.					×													
Tergitol TMN Series, X Series	Union Carbide Chemicals Co.						×												
Tetronic	Wyandotte Chemicals Corp.									×									
Texamid	Henkel International GmbH, A.H. Carnes															×			
Texapon	Henkel International GmbH, A.H. Carnes	×																	
T-Mulz	Thompson-Hayward Chemical Co.									×									
Triple C	Pro-Chem, Inc.			×															
Triton B1956	Rohm and Haas Co.									×									
Triton CF 10, 21, N-57 to 128, X-15 to 305*	Rohm and Haas Co.						×												
Triton GR	Rohm and Haas Co.		×																
Triton X151, 171, X152-172*	Rohm and Haas Co.					×													
Triton X200, X202	Rohm and Haas Co.				×														
Triton X301, 770, W-30	Rohm and Haas Co.	×																	

* Series of similar products.

Name	Supplier	Anionic					Nonionic				Cationic			Natural					
		Sulfates	Alkyl Sulfonic Acids and Their Salts	Carboxylic Acids and Their Salts	Alkylaryl Sulfonic Acids and Salts	Miscellaneous	Polyalkoxy Ethers	Alkoxy Esters	Alkanolamides	Miscellaneous	Amines and Amine Salts	Quaternary Amines and Their Salts	Miscellaneous	Amphoteric	Cellulose Derivatives	Water Soluble Gums	Lipids	Fluorocarbons	Silanes and Silicones
Triton X400	Rohm and Haas Co.											×							
Triton X700, X800	Rohm and Haas Co.					×													
Tween 20-85, 100, 280 VS	Atlas Chemical Industries, Inc.						×												
Vantoc N	I.C.I. Organics, Inc.											×							
Varamide Al	Varney Chemical Corp.							×											
Variquat K300	Varney Chemical Corp.											×							
Varonic NK	Varney Chemical Corp.						×												
Viscolan	American Cholesterol Products, Inc.							×											
Viscontran	Henkel International GmbH, A.H. Canes Co.														×				
Volpo 3, 10, 20	Croda, Inc.						×												
Waxolan	American Cholesterol Products, Inc.							×											
X-L Charge Detergent Honey	Dixo Company, Inc.					×													
Zyrol	Procter and Gamble, Co.			×															

Suppliers of Emulsifying Agents

1. Aceto Chemical Co.
2. Air Reduction Co., Inc.
3. Alabama Binder and Chemical Corp.
4. Alco Chemical Co.
5. Alcolac Chemical Corp.
6. Amalgamated Chemical Corp.
7. American Can Co., Marathon Products Division
8. American Cholesterol Products, Inc.
9. American Cyanamid Co.
10. American Dyewood, Inc.
11. American Lecithin Co.
12. Amoco Chemicals Corp.
13. Apex Chemical Co., Inc.
14. Archer Daniels Midland Co.
15. Ardmore Chemical
16. Arizona Chemical Co.
17. Arlen Chemical
18. Armour Industrial Chemical Co.
19. Arol Chemical Products
20. Atlas Chemical Industries, Inc.
21. Atlas Refinery, Inc.
22. Baker Castor Oil Co.
23. Barrett Chemical
24. BASF Colors and Chemicals, Inc.
25. Beacon Chemical Industries, Inc.
26. Berkshire Color and Chemical Co.
27. Boler Petroleum Co.
28. Burkart - Schier Chemical Co.
29. Bryant Chemical
30. Bryton Chemical Co.
31. Calgon Corp.
32. California Chemical Co.
33. Canada Packers Ltd.
34. Cargill, Inc.
35. Chemical Corp.
36. Carlisle Chemical Works, Inc.
37. Central Soya Co.
38. Chapman Chemical Co.
39. Chemactants, Inc.
40. Ciba Chemical and Dye Co., Inc.
41. Cindet Chemicals, Inc.
42. Clintwood Chemical Co.
43. Clough Chemical Co., Ltd.
44. Colgate-Palmolive Co.
45. Colloidal Products Corp.
46. Colloids, Inc.
47. Colonial Sugars Co.
48. Commercial Solvents Corp.
49. Continental Chemical Co.
50. Crest Chemical Corp.

51. Croda, Inc.
52. Crown Chemical Corp.
53. Crown Zellerbach Corp.
54. Culver Chemical Co.
55. DePaul Chemical Co.
56. Dexter Chemical Corp.
57. Diamond Alkali Co.
58. Distillation Products Industries
59. Dixo Company, Inc.
60. C. B. Dolge Co.
61. Dominion Products, Inc.
62. The Dow Chemical Co.
63. Drew Chemical Corp.
64. DuBois Chemicals, Inc.
65. E. I. du Pont de Nemours and Co.
66. Eastern Color and Chemical Co.
67. Emery Industries, Inc.
68. Emkay Chemical Co.
69. Essential Chemicals Corp.
70. W. F. Fancourt Co.
71. Far-Best Corp.
72. Fine Laboratories, Inc.
73. Fine Organics, Inc.
74. Finetex, Inc.
75. R. E. Flatow and Co., Inc.
76. Foremost Chemical Products Co.
77. Geigy Industrial Chemicals
78. General Aniline and Film Cerp.
79. General Mills
80. Georgia-Pacific Corp.
81. Goldschmidt Chemical Corp.
82. Glidden Co.
83. Glyco Chemicals, Inc.
84. B. F. Goodrich Chemical Co.
85. Greenwood Textile Supply Co.
86. Guardian Chemical Corp.
87. C. P. Hall Co.
88. Hart Products Corp.
89. Henkel International Gmbh, A. H. Carnes Co., agent
90. Hercules Powder Co.
91. Hexagon Laboratories Inc.
92. Hodag Chemical Corp.
93. Hooker Chemical Corp.
94. Hope Chemical
95. E. F. Houghton and Co.
96. Humble Oil and Refining Co.
97. I. C. I. Organics, Inc.
98. International Selling
99. Intex Chemical Corp.
100. Ion acChemical Co.
101. Isochem Corp.
102. Jefferson Chemical Co.
103. Andrew Jergens
104. Jersey State Chemicals
105. W. H. and F. Jordan, Jr., Mfg. Co.
106. Kali Mfg. Co.
107. Kalide Corp.
108. Kehew-Gradley and Co.
109. Kessler Chemical Co., Inc.
110. Knapp Products, Inc.
111. H. Kohnstamm and Co.
112. Laurel Soap Mfg.
113. Leatex Chemical Co.
114. Lever Brothers Co.
115. Leyda Oil and Chemical Co.
116. Maher Color and Chemical Co.
117. Malmstrom Color and

Chemical Corp.
118. Marden-Wild Corp.
119. Mathe Chemical Co.
120. Merix Chemical Co.
121. Metro-Atlantic, Inc.
122. M. Michel and Co.
123. Harry Miller Corp.
124. 3M Company
125. Miranol Chemical Co., Inc.
126. Mona Industries, Inc.
127. Monsanto Chemical Co.
128. Moretex Chemical Products
129. Murphy-Phoenix Oil Co.
130. National Lead Co.
131. Nopco Chemical Co.
132. Northwestern Chemical Co.
133. Nostrip Chemical Works, Inc.
134. Onyx Chemical Corp.
135. Ottol Oil Co.
136. Patent Chemicals
137. Pecks Products, Co.
138. Pennsalt Chemicals Corp.
139. Pennsylvania Refining Co.
140. Perry Brothers, Inc.
141. Pilot Chemical Co.
142. Charles Pfizer and Co.
143. Pro-Chem. Inc.
144. Proctor and Gamble, Co.
145. Proctor Chemical Co.
146. Proven Products
147. Rohm and Haas Co.
148. Relly-Whiteman-Walton Co.
149. Retzloff Chemical Co.
150. Richardson Co.
151. Robeco Chemicals, Inc.
152. Robinson Wagner Co., Inc.
153. Rozilda Laboratories
154. Ryco, Inc.
155. Sher Brothers
156. Scholler Brothers, Inc.
157. Seaboard Chemicals, Inc.
158. Shawinigan Resins Corp.
159. Shell Oil Co.
160. George F. Siddall, Co.
161. Werner G. Smith, Inc.
162. Sole Chemical Corp.
163. Solvol Chemical Co., Inc.
164. Sonneborn Chemical and Refining
165. Southern Sizing Co.
166. Fredrick A. Stresen-Reuter, Inc.
167. A. E. Staley Manufacturing Co.
168. Standard Chemical Co.
169. Standard Chemical Products
170. Stauffer Chemical Co.
171. Stepan Chemical Co.
172. Sun Chemical Corp.
173. Suffact Co., Inc.
174. Swift and Co.
175. Synthetic Chemicals, Inc.
176. Synthron, Inc.
177. Tanatex Chemical Corp.
178. Textilana Corp.
179. Textron, Inc.
180. Thompson Chemical Corp.
181. Thompson-Hayward Chemical Co.
182. Titan Chemical Products, Inc.
183. Arthur C. Trask Co.
184. Treplow Chemicals, Inc.
185. Trylon Chemicals, Inc.
186. Union Carbide Corporation (Chemicals and Silicones Divisions)
187. United Merchants and

Manufacturers, Inc.
188. Universal Chemicals Corp.
189. Van Dyk and Co.
190. Varney Chemical Corp.
191. Verona-Pharma Chemical Corp.
192. Wasco Laboratories
193. Washine Chemical Corp.
194. Wayland Chemical Corp.
195. Wilson and Co., Inc.
196. Witco Chemical Co., Inc.
197. W. A. Wood, Co.
198. Woonsocket Color and Chemical Co.
199. Wyandotte Chemicals Corp.
200. General Electric Silicones Division
201. Dow Corning Corp.
202. Allied Chemical Corp, National Aniline Division
203. Chevron Chemical Co., Oronite Division
204. Tennessee Corp.

INDEX